Essential
Cell Biology
Test Bank

Essential
Cell Biology
Test Bank

Kelly Komachi

Rosann Tung

Garland Publishing, Inc.

New York & London

Kelly Komachi is a molecular biologist at the University of California at San Francisco. She is responsible for the creation of questions for Chapters 1 through 15 and 17.

Rosann Tung is a cell biologist at Harvard Medical School and is responsible for the creation of questions for Chapters 16, 18, and 19.

© 1997 by Kelly Komachi and Rosann Tung

CIP information is available.

ISBN: 0-8153-2778-1

Published by Garland Publishing, Inc.

717 Fifth Avenue, New York, NY 10022

Printed in the United States of America

Preface

The questions in this test bank are arranged according to the sections in *Essential Cell Biology* to which they mainly refer.

Questions are graded into three levels of difficulty: Easy, Intermediate, and Difficult. Easy questions are mainly factual recall. Some of these are addressed to relatively small points and should be used only if these points have been emphasized in your teaching. Intermediate and Difficult questions require more reasoning on the part of the student. The Difficult questions are quite challenging, and four or five of them may be sufficient for a one-hour exam.

Questions are also labeled as one of five types: multiple choice (please note that multiple choice questions may have more than one correct answer), short answer, matching/fill in blanks, data interpretation, and art labeling.

Contents

Essential
Cell Biology
Test Bank

1 Introduction to Cells

Questions

CELLS UNDER THE MICROSCOPE (Pages 1–9)

1–1 Easy, multiple choice

The smallest unit of life is:

    A. a DNA molecule.

    B. a cell.

    C. an organelle.

    D. a virus.

    E. a protein.

The Invention of the Light Microscope Led to the Discovery of Cells (Pages 2–3)
Cells, Organelles, and Even Molecules Can Be Seen Under the Microscope (Pages 3–9)

1–2 Easy, multiple choice

What unit of length would you generally use to give the measurements of a typical human cell?

    A. Centimeters.

    B. Nanometers.

    C. Millimeters.

    D. Micrometers.

1–3 Intermediate, short answer (Requires information from Panel 1–1)

(A) What sets the limit on the size of structure that can be seen in the light microscope?

(B) Why are tissues usually cut into thin sections and stained before examination under the light microscope?

1–4 Intermediate, multiple choice (Requires information from Panel 1–1)

For which of the following would you use an electron microscope rather than a light microscope?

    A. To pick out individual human cells from a cell culture in order to grow new cultures from them.

    B. To watch organelles move inside a cell.

    C. To look at cells stained with fluorescent dyes.

    D. To see ribosomes.

    E. To study moving bacteria.

THE EUCARYOTIC CELL (Pages 9–17)

1–5   Easy, multiple choice

The most reliable feature distinguishing a eucaryotic cell from a procaryotic cell is:

    A. the presence of a plasma membrane.

    B. the presence of a nucleus.

    C. the eukaryotic cell's larger size.

    D. the presence of internal membranes.

    E. the presence of DNA.

1–6   Intermediate, multiple choice (Requires information from section on pages 22–25)

Which of the following statements concerning procaryotes are true?

    A. They have no nucleus and hence no DNA.

    B. They have no Golgi apparatus.

    C. They can form simple multicellular organisms.

    D. They include bacteria, yeast, and protozoans.

    E. They are all able to live on inorganic energy sources.

The Nucleus Is the Information Store of the Cell (Pages 9–10)

1–7   Intermediate, multiple choice (Requires information from sections on pages 10–12 and 31–33, and Panel 1–3)

Which of the following statements are correct?

    A. Chromosomes are always visible in human cells in the light microscope.

    B. All the DNA in a eucaryotic cell is contained in the nucleus.

    C. A human white blood cell and a human nerve cell contain different numbers of chromosomes.

    D. A human sperm cell and a human white blood cell contain different numbers of chromosomes.

    E. All eucaryotic organisms have the same number of chromosomes in their cells.

Mitochondria Generate Energy from Food to Power the Cell (Pages 10–12)

1–8   Easy, short answer

Correct each of the following so that it becomes a true statement about mitochondria.

    A. Mitochondria take in carbon dioxide and release oxygen.

    B. ADP is synthesized from ATP in mitochondria.

    C. Mitochondria are enclosed by two membranes, the outer one of which is highly folded.

    D. Mitochondria are thought to be derived from photosynthetic bacteria.

    E. Mitochondria are found in aerobic procaryotes.

1–9   Intermediate, multiple choice (Requires information from sections on pages 12–13 and 25–26)

Which of the following require mitochondria in order to live?

    A. Eubacteria (Bacteria).

    B. Plant cells with chloroplasts.

    C. Archaebacteria (Archaea).

    D. The unicellular microorganism *Giardia*.

    E. None of the above.

1–10   Difficult, short answer

In an aerobic bacterium, where do you think most of the proteins responsible for cellular respiration are located?

Chloroplasts Capture Energy from Sunlight (Pages 12–13)

1–11   Easy, multiple choice

Which one or more of the following statements is true for mitochondria only, and not for both mitochondria and chloroplasts?

    A. They are enclosed by a double membrane.

    B. They are thought to be derived from procaryotes.

    C. They cannot grow and reproduce when isolated from the cell.

    D. They reproduce by dividing in two.

    E. They are found in all aerobic eucaryotic cells.

1–12   Difficult, short answer

You fertilize egg cells from a healthy plant with pollen (which contains the male germ cells) that has been treated with DNA-damaging agents. You find that some of the offspring have defective chloroplasts, and that this characteristic can be passed on to future generations. This surprises you at first because you happen to know that the male germ cell in the pollen grain contributes no chloroplasts to the fertilized egg cell and thus to the offspring. What can you deduce from these results?

Internal Membranes Create Intracellular Compartments with Different Functions (Pages 13–15)

1–13   Easy, multiple choice

Which of the following organelles are surrounded by two layers of membrane?

    A. Endoplasmic reticulum.
    B. Nucleus.
    C. Lysosome.
    D. Peroxisome.
    E. Vacuole.

1–14   Intermediate, short answer

In a eucaryotic cell specialized for secretion, which internal organelles would you expect to be particularly abundant?

1–15   Intermediate, multiple choice

A white blood cell encounters a bacterium and engulfs it. Which organelle will the bacterium end up in before it is destroyed?

    A. Lysosome.
    B. Golgi apparatus.
    C. Endoplasmic reticulum.
    D. Nucleus.
    E. Peroxisome.

The Cytosol Is a Concentrated Aqueous Gel of Large and Small Molecules (Pages 15–16)

1–16   Easy, short answer

From the list below select the THREE cellular structures or compartments that are found in all cells.

    A. Nucleus.
    B. Ribosomes.
    C. Cytosol.
    D. Mitochondria.
    E. Chloroplasts.
    F. Plasma membrane.
    G. Endoplasmic reticulum.
    H. Lysosomes.

The Cytoskeleton Is Responsible for Cell Movements (Pages 16–17)

1–17   Easy, multiple choice

Which of the following statements regarding the cytoskeleton are false?

    A. It is made up of actin filaments, microtubules, and intermediate filaments.

    B. It is involved in cell movement and thus is absent from plant cells.

    C. It is involved in muscle contraction.

    D. It is being continually rearranged.

    E. It is anchored to the plasma membrane.

1–18   Intermediate, multiple choice (Requires information from Panel 1–1)

You have grown a culture of human cells and discover that they are heavily contaminated with bacteria. Which of the following procedures is most likely to eliminate the bacteria without killing the human cells?

    A. Treating the culture with a drug that causes microtubules to fall apart.

    B. Diluting a small portion of the contaminated culture with 1000 times as much fresh nutrient broth and regrowing the cells.

    C. Treating the culture with a drug that damages DNA.

    D. Treating the culture with a drug that dissolves cell walls.

    E. Treating the culture with a detergent that destroys cell membranes.

1–19   Easy, art labeling (Requires information from sections on pages 9–17)

On the schematic drawing of an animal cell in Figure Q1–19 match the labels given in the list below to the numbered label lines.

    A. Plasma membrane.

    B. Nuclear envelope.

    C. Cytosol.

    D. Golgi apparatus.

    E. Endoplasmic reticulum.

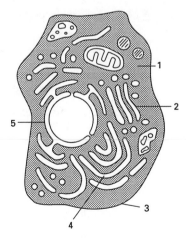

Q1–19

UNITY AND DIVERSITY OF CELLS (Pages 17–33)
Cells Vary Enormously in Appearance and Function (Pages 19–21)

1–20   Intermediate, multiple choice (Requires student to have studied the whole chapter)

Which of the following statements are correct?

    A. Snakes are larger than worms because their cells are larger.
    B. All the cells in a mouse are the same size.
    C. Even the largest cells are not visible to the naked eye.
    D. Some bacterial cells are as large as a yeast cell.
    E. Procaryotic cells are generally larger than eucaryotic cells.

Living Cells All Have a Similar Basic Chemistry (Page 21)

1–21   Easy, multiple choice

Which of the following statements about the basic chemistry of cells are true?

    A. All cells contain exactly the same proteins.
    B. All proteins are constructed from the same 22 amino acids.
    C. All genetic instructions in cells are stored in DNA.
    D. All organisms contain the same genes.
    E. All of the above.

All Present-Day Cells Have Apparently Evolved from the Same Ancestor (Pages 21–22)

1–22   Intermediate, multiple choice

Which of the following statements about the evolution of present-day organisms are true?

    A. Eucaryotes preceded procaryotes.
    B. Eucaryotes arose from eubacteria long after archaebacteria and eubacteria had diverged from each other.
    C. Eucaryotes acquired chloroplasts before they acquired mitochondria.
    D. Eucaryotes acquired mitochondria before they acquired chloroplasts.
    E. The common ancestor cell was an archaebacterium.

1–23   Intermediate, multiple choice

Which of the following statements are true?

    A. Mutations are always harmful to an organism.
    B. Mutation is essential for evolution to occur.
    C. Mutation is the only source of genetic differences between parents and off-spring in plants and animals.

D. Mutation always leads to evolution.

E. Mutations always lead to evolutionary "dead ends."

Bacteria Are the Smallest and Simplest Cells (Pages 22–25)

1–24 Intermediate, multiple choice

Bacteria can evolve faster than animals chiefly because:

A. they reproduce more frequently.

B. they can carry out a much wider range of chemical reactions.

C. their cells are simpler.

D. they inhabit a wide range of habitats.

E. they possess cell walls.

Molecular Biologists Have Focused on *E. coli* (Page 25)

1–25 Easy, short answer

Why is information on basic cell processes obtained from the bacterium *E. coli* helpful in understanding eucaryotic cells as well?

***Giardia* May Represent an Intermediate Stage in the Evolution of Eucaryotic Cells (Pages 25–26)**

1–26 Intermediate, short answer

Why is the unicellular microorganism *Giardia* thought to represent an intermediate stage in the evolution of eucaryotic cells?

Brewer's Yeast Is a Simple Eucaryotic Cell (Pages 26–27)

1–27 Easy, short answer

(A) In what way does a fungal cell structurally resemble a plant cell rather than an animal cell?

(B) Which principal organelle does a plant cell contain that a fungal cell does not?

Single-celled Organisms Can Be Large, Complex, and Fierce: The Protozoans (Pages 27–28)

1–28 Intermediate, short answer

The protozoan *Didinium* feeds on other organisms by engulfing them. Why are bacteria in general unable to feed on other cells in this way?

The World of Animals Is Represented by a Fly, a Worm, a Mouse, and *Homo Sapiens* (Pages 29–31)

1–29   Intermediate, short answer

You wish to explore how mutations in specific genes affecting sugar metabolism might alter tooth development. Which organism is likely to provide the best model system for your studies, and why?

- A. Humans.
- B. Mice.
- C. Chipmunks.
- D. Sharks.
- E. Elephants.

Cells in the Same Multicellular Organism Can Be Spectacularly Different (Pages 31–33)

1–30   Easy, multiple choice

The specialized cell types in the body of a multicellular organism are different from each other chiefly because:

- A. each cell type contains different genes.
- B. different genes are switched on in different cell types.
- C. some cell types have lost some of the genes that were present in the fertilized egg.
- D. the fertilized egg divides by cell divisions that do not give rise to genetically identical cells.
- E. the different cell types contain fundamentally different organelles.

1–31   Intermediate, short answer (Requires student to have studied the whole chapter)

List the following items in order of size from the smallest to the largest.

- A. Protein molecule.
- B. Human fat cell.
- C. Carbon atom.
- D. Ribosome.
- E. Yeast cell.
- F. Mitochondrion.

1–32   Intermediate, matching/fill in blanks (Requires student to have studied the whole chapter)

Match each of the cells in the first list with its description selected from the second list by writing the appropriate number beside it.

    A. Neuron.

    B. Schwann cell.

    C. Lymphocyte.

    D. Fibroblast.

    E. Adipose cell.

1. Connective tissue cell specialized to secrete fibrous extracellular matrix.
2. Nervous system cell specialized for electrical signaling.
3. Blood cell specialized to ingest bacteria.
4. Nervous system cell specialized for insulation.
5. Connective tissue cell specialized to secrete bone matrix.
6. Connective tissue cell specialized to store fat.
7. Blood cell involved in production of antibodies for defense against infection.
8. Nervous system cell specialized to detect external stimuli.

Answers

A1–1.   B.

A1–2.   D.

A1–3.   (A) The wavelength of visible light. (B) Most tissues are not transparent enough to be examined directly in the light microscope. Transparency is increased by slicing them into thin sections, and staining shows up the different cellular structures in contrasting colors.

A1–4.   D.

A1–5.   B.

A1–6.   B and C.

A1–7.   D.

A1–8.   A. Mitochondria take in <u>oxygen</u> and release <u>carbon dioxide</u>.
        B. <u>ATP</u> is synthesized from <u>ADP</u> in mitochondria.
        C. Mitochondria are enclosed by two membranes, the <u>inner</u> one of which is highly folded.
        D. Mitochondria are thought to be derived from <u>aerobic</u> bacteria.
        E. Mitochondria are found in aerobic <u>eucaryotes</u>.

A1–9.   B.

A1–10. In the plasma membrane. According to the theory of mitochondrial origin outlined in this chapter, the plasma membrane of the engulfed bacterium would have become the inner mitochondrial membrane, where most of the proteins involved in cellular respiration are located.

A1–11. E.

A1–12. Your results show that not all of the information required for making a chloroplast is encoded in the chloroplast's own DNA, but that some at least must be encoded in the DNA carried in the nucleus.

   The reasoning is as follows. Genetic information is only carried in DNA and thus the defect in the chloroplasts must be due to a mutation in DNA. But all of the chloroplasts in the offspring (and thus all of the chloroplast DNA) must derive from those in the female egg cell, since chloroplasts only arise from other chloroplasts. Hence, all of the chloroplasts contain undamaged DNA from the female parent's chloroplasts. In all the cells of the offspring, however, half of the nuclear DNA will have come from the male germ cell nucleus, which combined with the female egg nucleus at fertilization. Since this DNA has been treated with DNA-damaging agents, it must be the source of the heritable chloroplast defect. Thus some of the information required for making a chloroplast is encoded by the nuclear DNA.

A1–13. B.

A1–14. The endoplasmic reticulum and the Golgi apparatus.

A1–15. A.

A1–16. B, C, and F.

A1–17. B.

A1–18. D. Bacteria have cell walls whereas mammalian cells do not.

A1–19. 1, C; 2, D; 3, A; 4, E; 5, B.

A1–20. D.

A1–21. C.

A1–22. D.

A1–23. B.

A1–24. A.

A1–25. Because of their common evolutionary origin, bacteria and eucaryotes perform the most basic cell processes in essentially the same way. These include the use of DNA as the store of genetic information, its replication, and how it is read out to direct the synthesis of proteins.

A1–26. It possesses a nucleus and cytoskeleton, which are diagnostic of eukaryotic cells, but it does not possess mitochondria or chloroplasts, which are thought to have been acquired by eukaryotic cells later. DNA analysis also shows that *Giardia* diverged from other eukaryotic cells very early in their evolution.

A1–27. (A) Like plant cells, fungal cells have cell walls. (B) Chloroplasts.

A1–28. *Didinium* engulfs prey by changing its shape, and for this it uses its cytoskeleton. Bacteria have no cytoskeleton, and cannot easily change their shape because they are generally surrounded by a tough cell wall.

A1–29. D. Chipmunks, sharks, and elephants all have remarkable teeth but largely unknown genetics; they offer no practicable way to obtain mutations in the genes of interest. Humans will be more informative, since many naturally occurring mutations are known, including some that affect sugar metabolism. New mutations, however, cannot be created to order: it would be ethically unacceptable. Mice are probably the best option: their genetics and biology are well studied, and they can now be bred with purposely engineered mutations in specific genes.

A1–30. B.

A1–31. C, A, D, F, E, B.

A1–32. A, 2; B, 4; C, 7; D, 1; E, 6.

2 Chemical Components of Cells

Questions

CHEMICAL BONDS (Pages 37–52)
Cells Are Made of Relatively Few Types of Atoms (Pages 38–39)

2–1 Easy, multiple choice

Which of the following are the same in all atoms of an element?

    A. Number of neutrons.
    B. Number of protons.
    C. Mass.
    D. Atomic weight.
    E. Number of neutrons plus protons.

2–2 Easy, short answer

If the isotope ^{32}S has 16 protons and 16 neutrons, how many protons and how many neutrons will the isotope ^{37}S have?

2–3 Easy, short answer

(A) If 0.5 mole of glucose weighs 90 g, what is the molecular weight of glucose?

(B) What is the concentration in grams per liter (g/l) of a 0.25 M solution of glucose?

(C) How many molecules are there in 1 mole of glucose?

2–4 Easy, short answer

Which of the following elements is least abundant in living organisms?

    A. Sulfur.
    B. Carbon.
    C. Oxygen.
    D. Nitrogen.
    E. Hydrogen.

The Outermost Electrons Determine How Atoms Interact (Pages 39–42)

2–5 Easy, multiple choice

Atoms form covalent bonds with each other by:

    A. sharing protons.
    B. sharing electrons.
    C. transferring electrons from one atom to the other.
    D. sharing neutrons.
    E. attraction of positive and negative charges.

Ionic Bonds Form by the Gain and Loss of Electrons (Pages 42–43)

2–6 Easy, multiple choice

An ionic bond between two atoms is formed as a result of:

    A. sharing of electrons.
    B. loss of a neutron from one atom.
    C. loss of electrons from both atoms.
    D. loss of a proton from one atom.
    E. transfer of electrons from one atom to the other.

2–7 Easy, data interpretation

Which of the following pairs of elements are likely to form ionic bonds? Use Figure Q2–7 if necessary.

    A. Hydrogen and hydrogen.
    B. Magnesium and chlorine.
    C. Carbon and oxygen.
    D. Sulfur and hydrogen.
    E. Carbon and chlorine.

Q2–7

Covalent Bonds Form by the Sharing of Electrons (Pages 43–45)

2–8 Intermediate, short answer

(A) In which scientific units is the strength of a chemical bond usually expressed?

(B) If 0.5 kilocalories of energy is required to break 6×10^{23} bonds of a particular type, what is the strength of this bond?

There Are Different Types of Covalent Bonds (Pages 45–48)

2–9   Easy, matching/fill in blanks (Requires information from sections on pages 42–48)

For each of the following sentences, fill in the blanks with the correct word selected from the list below. Use each word only once.

    A. A molecule is a cluster of atoms held together by _____ bonds.

    B. An ionic bond is an example of a _____ bond.

    C. The magnesium and chlorine atoms in magnesium chloride are held together by _____ bonds.

    D. Covalent bonds are individually _____ than noncovalent bonds.

weaker; ionic; stronger; nonpolar; covalent; hydrophobic; hydrogen; noncovalent.

2–10   Intermediate, data interpretation

After looking at Figure Q2–7 above, which of the following pairs of atoms do you expect to be able to form double bonds with each other?

    A. Mg and Ca.

    B. C and Cl.

    C. S and O.

    D. C and H.

    E. He and O.

Water Is the Most Abundant Substance in Cells (Pages 48–49)

2–11   Intermediate, short answer (Requires information from Panel 2–2)

(A) Sketch three different ways three water molecules could be held together by hydrogen bonding.

(B) On a sketch of a single water molecule, indicate the distribution of positive and negative charge (using the symbols δ^+ and δ^-).

(C) How many hydrogen bonds can a hydrogen atom in a water molecule form? How many hydrogen bonds can the oxygen atom in a water molecule form?

2–12   Intermediate, multiple choice

Which of the following statements about hydrogen bonds are true?

    A. They are weak covalent bonds that are easily disrupted by heat

    B. They are weak bonds formed between hydrocarbons in water.

    C. They are weak bonds formed between nonpolar groups.

    D. They are weak bonds only formed in the presence of water.

    E. They are weak bonds involved in maintaining the 3-D structure of macromolecules.

2–13   Intermediate, multiple choice (Requires information from Panel 2–2)

Based on what you know about the properties of water, which of the following statements about methanol (CH_3OH) are true?

    A. Methanol molecules form more hydrogen bonds than water molecules do.

    B. The boiling point of methanol is higher than that of water.

    C. Salts such as NaCl are less soluble in methanol than in water.

    D. Methanol is a more cohesive liquid than water.

    E. Methanol has a higher surface tension than water.

Some Polar Molecules Form Acids and Bases in Water (Pages 49–52)

2–14   Intermediate, short answer (Requires information from Panel 2–2)

(A) What is the pH of pure water?

(B) What concentration of hydronium ions does a solution of pH 8 contain?

(C) Complete the following reaction:

    $CH_3COOH + H_2O \rightarrow$

(D) Will the reaction in C occur more readily if the pH of the solution is high or if it is low?

MOLECULES IN CELLS (Pages 52–73)
A Cell Is Formed from Carbon Compounds (Page 52)

2–15   Intermediate, multiple choice (Requires information from sections on pages 37–49 and Panel 2–1)

Carbon atoms CANNOT:

    A. form four covalent bonds to other atoms.

    B. form both covalent and ionic bonds.

    C. form double bonds.

    D. form chains of virtually unlimited size.

    E. form ring structures.

2–16 Easy, matching/fill in blanks

Match the chemical groups shown in the first list with their names selected from the second list.

	List 1		List 2
A.	–OH	1.	Amino.
B.	–C=O	2.	Aldehyde.
C.	–COOH	3.	Phosphate.
D.	–CH$_3$	4.	Carboxyl.
E.	–NH$_2$	5.	Carbonyl (ketone).
		6.	Methyl.
		7.	Amido.
		8.	Ester.
		9.	Hydroxyl.

Cells Contain Four Major Families of Small Organic Molecules (Pages 52–53)

2–17 Intermediate, multiple choice (Requires information from sections on pages 53–55 and 61–65, and Panels 2–3, 2–4, and 2–6)

Which of the following small molecules contain nitrogen?

A. Guanine.

B. Glycerol.

C. Stearic acid.

D. Ribose.

E. Mannose.

Sugars Are Energy Sources for Cells and Subunits of Polysaccharides (Pages 53–55)

2–18 Intermediate, multiple choice (Requires information from Panel 2–5)

Which of the following are examples of isomers?

A. ^{14}C and ^{12}C.

B. Alanine and glycine.

C. Adenine and guanine.

D. Glycogen and cellulose.

E. Glucose and galactose.

2–19 Intermediate, multiple choice (Requires information from Panel 2–3)

Cellulose and starch are both polysaccharides composed entirely of glucose monomers, yet one forms fibers and the other forms granules. Which of the following could NOT help to account for these differences?

A. Linkage of some monomers to more than one other monomer.

B. Differences in the lengths of the polysaccharide chains.

C. The linkage of some monomers through chemical groups other than hydroxyl groups.

D. The linkage of monomers through α links in starch and β links in cellulose.

E. The 3-D structures of the polysaccharide molecules.

Fatty Acids Are Components of Cell Membranes (Pages 55–60)

2–20   Easy, data interpretation

(A) How many carbon atoms does the molecule represented in Figure Q2–20 have?

(B) How many hydrogen atoms?

(C) What type of molecule is it?

Q2–20

2–21   Intermediate, art labeling

On the phospholipid molecule in Figure Q2–21 match the correct labels selected from the list below to the numbered label lines.

A. Phosphate.

B. Nonpolar headgroup.

C. Glycerol.

D. Polar head group.

E. Fatty acid.

F. Acetic acid.

G. Sugar.

H. Hydrophobic region.

I. Hydrophilic region.

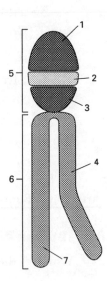

Q2–21

2–22   Easy, multiple choice

Phospholipids can form bilayer membranes because they are:

A. hydrophobic.

B. lipids.

C. amphipathic.

D. hydrophilic.

E. amphoteric.

2–23   Easy, multiple choice

Which of the following statements regarding lipids are false?

    A. Lipids can serve as energy sources for cells.

    B. All cell membranes contain lipids.

    C. All cells contain lipids.

    D. All lipids can form bilayer membranes.

    E. Lipids can function as hormones.

Amino Acids Are the Subunits of Proteins (Pages 60–61)

2–24   Intermediate, short answer (Requires information from Panel 2–5)

(A) Write out the sequence of amino acids in the following peptide using the full names of the amino acids.

Pro—Val—Thr—Gly—Lys—Cys—Glu

(B) According to the conventional way of writing the sequence of a peptide or a protein, which is the C-terminal amino acid and which the N-terminal amino acid in the above peptide?

2–25   Intermediate, multiple choice (Requires information from Panel 2–5)

Which of the following statements about amino acids are true?

    A. Twenty-two amino acids are commonly found in proteins.

    B. Most of the amino acids used in protein biosynthesis have charged side chains.

    C. Not all amino acids have stereoisomers.

    D. D- and L-amino acids are found in proteins.

    E. All amino acids contain an NH_2 and a COOH group.

2–26   Intermediate, multiple choice (Requires information from Panel 2–1)

Proteins can be modified by reaction with acetate which results in the addition of an acetyl group to lysine side chains as shown in Figure Q2–26.

The bond indicated in the acetylated lysine side chain is most like:

    A. an ester.

    B. a peptide bond.

    C. a phosphoanhydride bond.

    D. a hydrogen bond.

    E. a phosphoester bond.

Q2–26

Nucleotides Are the Subunits of DNA and RNA (Pages 61–65)

2–27   Easy, multiple choice

DNA differs from RNA in:

    A. the number of different bases used.

    B. the number of phosphates between the sugars in the sugar-phosphate back-bone.

    C. one of the pyrimidines used.

    D. one of the purines used.

    E. the chemical polarity of the polynucleotide chain.

2–28   Intermediate, matching/fill in blanks (Requires student to have studied the whole chapter)

From list A below, which one reaction type describes ALL of the following reactions?

ADP $\rightarrow$ ATP

Fatty acids + glycerol $\rightarrow$ triacylglycerol

Glucose $\rightarrow$ glycogen

Deoxynucleotides $\rightarrow$ DNA

List A: phosphorylation; esterification; condensation; hydrolysis; polymerization.

2–29   Intermediate, multiple choice (Requires student to have studied the whole chapter)

Which of the following are likely to be disrupted by high concentrations of salt?

    A. A lipid bilayer.

    B. The peptide bonds in a protein.

    C. A complex of two proteins.

    D. The sugar-phosphate backbone of a nucleic acid.

    E. An oil droplet in water.

Macromolecules Contain a Specific Sequence of Subunits (Pages 65–69)

2–30   Intermediate, short answer

You are trying to make a synthetic copy of a particular protein but accidentally join the amino acids together in exactly the reverse order. One of your classmates says the two proteins must be identical, and bets you $20 that your synthetic protein will have exactly the same biological activity as the original. After having read this chapter, you have no hesitation in staking your $20 that it won't. What particular feature of a polypeptide chain makes you sure your $20 is safe, but that your project will have to be redone.

Noncovalent Bonds Specify the Precise Shape of a Macromolecule (Pages 69–72)

2–31   Intermediate/difficult, short answer

A protein chain folds into its three-dimensional structure by making many noncovalent bonds between different parts of the chain. Would the protein retain its shape if you cleaved all the peptide bonds between the amino acids after the protein had folded up? Explain your answer.

Noncovalent Bonds Allow a Macromolecule to Bind Other Selected Molecules (Pages 72–73)

2–32   Intermediate, multiple choice

DNA is negatively charged at physiological pH. A protein Z binds to DNA through noncovalent ionic interactions involving lysines. What will be the effect of acetylation of the lysine side chains (see Figure Q2–26) in protein Z on the strength of this binding:

    A. It should increase it because the acetylated lysine will form a greater number of ionic interactions with DNA.

    B. It should decrease it because the acetylated lysine no longer has a positive charge.

    C. It should have no effect because the unmodified lysine would not have formed an ionic interaction with the DNA.

    D. It should have no effect because the bond formed between lysine and the acetyl group still has a positive charge.

    E. It should decrease it unless the DNA can become more negatively charged.

2–33   Difficult, multiple choice + short answer

You are studying a microorganism in which a "male" turns pink in the presence of a "female." The male becomes pink because a protein X secreted by the female binds to and activates a protein Y on the male that is responsible for the color change. You have isolated a strain of the microorganism that produces a mutant form of protein X. This strain behaves normally at temperatures lower than 37°C, but at higher temperatures it cannot turn pink. Could any of the following changes in mutant protein X explain your results? If so, which ones, and explain why.

    A. It makes an extra hydrogen bond to protein Y.

    B. It makes fewer hydrogen bonds to protein Y.

    C. It makes a covalent bond to protein Y.

    D. It is completely unfolded at temperatures lower than 37°C.

    E. It is completely unfolded at temperatures higher than 37°C.

    F. It is unable to bind to protein Y at any temperature.

Answers

A2–1.   B. The number of neutrons, and hence the atomic weight and the mass, in atoms of a given element varies for different isotopes.

A2–2.   16 proteins and 21 neutrons.

A2–3.   (A) 180 daltons. A mole of a substance is equivalent to its molecular weight expressed in grams. (B) 45 g/l. (C) 6×10^{23}.

A2–4.   A. The most abundant elements are C, H, N, O.

A2–5.   B.

A2–6.   E.

A2–7.   B.

A2–8.   (A) kilocalories per mole (*or* kilojoules per mole). (B) 0.5 kcal/mole.

A2–9.   A. A molecule is a cluster of atoms held together by <u>covalent</u> bonds.
        B. An ionic bond is an example of a <u>noncovalent</u> bond.
        C. The magnesium and chlorine atoms in magnesium chloride are held together by <u>ionic</u> bonds.
        D. Covalent bonds are individually <u>stronger</u> than noncovalent bonds.

A2–10. C. Sulfur and oxygen both require two electrons to fill their outer shell and can do so by sharing four electrons and forming a double bond.

A2–11. (A) Figure A2–11a. (B) Figure A2–11b. (C) hydrogen can form one, oxygen two.

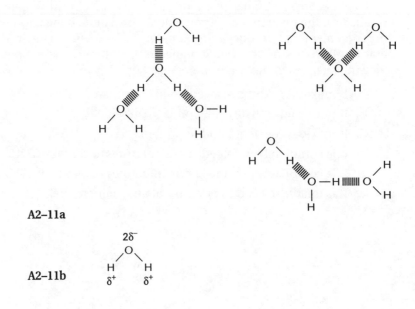

A2–11a

A2–11b

A2–12. E. A is false because hydrogen bonds are not covalent. B is false because the nonpolar –CH groups on hydrocarbons cannot form good hydrogen bonds, in water or out of it. C is essentially another way of stating B and thus is false. D is false because many molecules besides water can form hydrogen bonds and do so regardless of whether or not water is present.

A2–13. C. In methanol one of the hydrogens of a water molecule has been replaced by a nonpolar methyl group. Methanol will form fewer hydrogen bonds (thus A is false) and make fewer ionic interactions than water does. The ability of water to dissolve salts is a direct consequence of its ability to make ionic interactions. Salts are therefore less soluble in methanol. B, D, and E are all false since the high boiling point, high degree of cohesion and high surface tension of water are all a result of the extensive hydrogen bonding betwen water molecules. Since methanol makes fewer hydrogen bonds, its boiling point will be lower, it will be less cohesive, and it will have a lower surface tension than water.

A2–14. (A) pH 7. (B) 10^{-8} M. (C) $CH_3COO^- + H_3O^+$. (D) If the pH is high.

A2–15. B. Carbon atoms do not readily ionize, and both C^+ and C^- are extremely unstable.

A2–16. A, 9; B, 5; C, 4; D, 6; E, 1.

A2–17. A. Guanine is one of the nitrogenous bases in DNA and RNA. Ribose and mannose are carbohydrates; glycerol is a sugar alcohol and stearic acid is a fatty acid.

A2–18. E. Glucose and galactose are both six-carbon sugars and thus both have the formula $C_6H_{12}O_6$. They are thus isomers of each other. ^{14}C and ^{12}C are examples of isotopes. Adenine and guanine are bases containing different numbers of nitrogen and oxygen atoms. Glycogen and cellulose are different polymers of glucose. Alanine and glycine are amino acids with quite different side chains, a methyl group and a hydrogen atom, respectively.

A2–19. C. All the glucose monomers in these polysaccharides are linked through hydroxyl groups. A, B, and D could all in principle affect the 3-D structure of a polysaccharide chain and thus its physical properties. In fact, starch is a mixture of long branched chains and shorter unbranched helical chains in which the linkages are all α. Cellulose is formed of straight, unbranched chains in which the linkages are all β.

A2–20. (A) 20 carbon atoms. (B) 31 hydrogen atoms. (C) A fatty acid (Figure A2–20—it is in fact arachidonic acid).

A2–20

A2–21. 1, D; 2, A; 3, C; 4, E; 5, I; 6, H; 7, E.

A2–22. C.

A2–23. D.

A2–24.   (A) proline-valine-threonine-glycine-lysine-cysteine-glutamic acid (*or* glutamate). (B) C-terminal is glutamic acid (*or* glutamate); N-terminal is proline.

A2–25.   C. Glycine has no stereoisomers. A is false as there are only 20 amino acids. B is false because only four or five amino acids out of the 20 are charged at physiological pH. D is false because only L-amino acids are found in proteins. E is false since proline contains an NH but not an NH_2 group.

A2–26.   B. Like a peptide bond, the indicated bond is formed by reaction between a carboxyl group and an amino group.

A2–27.   C. RNA contains the pyrimidine uracil, while DNA contains the pyrimidine thymine. All the other features are the same in RNA and DNA.

A2–28.   Condensation.

A2–29.   C. Noncovalent ionic interactions such as those that hold two proteins together are most likely to be disrupted by salt. Lipid bilayers (A) and a lipid droplet (E) are held together by "hydrophobic interactions" on which salt will have no effect, and the other two options (B and D) are examples of covalent bonds, which are not disrupted by salt.

A2–30.   Because a peptide bond ($-\overset{\overset{H}{|}}{N}-\overset{\overset{O}{\|}}{C}-$) has a distinct chemical polarity, a polypeptide chain also has a distinct polarity. Therefore, the reversed protein chains cannot make the same noncovalent interactions as they fold up and thus will not adopt identical 3-D structures. Since the activity of a protein depends on its 3-D structure, the activities of these two proteins will definitely be different; in fact it is very unlikely that the reverse chain will fold into a defined—and hence functionally useful—structure at all, as it has not passed the stringent selective pressures imposed during evolution.

A2–31.   The protein would not retain its shape; it would simply fall apart into its original individual amino acids. A protein molecule is held together by the strong peptide bonds between the amino acids, not by the weak, noncovalent forces that govern folding.

A2–32.   B. Unmodified lysine side chains are positively charged and hence attractive to the negatively charged DNA (thus C is incorrect). Because acetylation neutralizes the positive charge (thus D is incorrect), the acetylated form of protein Z will form fewer ionic bonds with DNA (thus A is incorrect), and thus the strength of the interaction will decrease. E is incorrect, since increasing the number of negative charges on DNA would have no effect once the positive charge on the lysine has been neutralized.

A2–33.   B and E are possible explanations. If protein X makes fewer hydrogen bonds to protein Y, the two proteins will bind less tightly and may come apart at temperatures above 37°C, since thermal motion is one of the forces that disrupts weak bonds. The male will therefore not be able to turn pink above 37°C. If protein X is completely unfolded it will not be able to bind to protein Y, so explanation E could be the correct answer. In contrast, D would result in a protein X that is able to bind to protein Y only at high temperatures, and would result in a strain that would turn pink only at high temperatures. Explanation A would produce a protein that would bind more tightly than the normal protein X to protein Y and would therefore be likely to bind (and to turn pink) at temperatures above 37°C. If a covalent bond was made (C), it is unlikely that such a bond would be disrupted by any temperature that the microorganism could survive; the microorganism would therefore turn pink at any temperature. F would result in a strain that could not turn pink at any temperature.

3 Energy, Catalysis, and Biosynthesis

Questions

CATALYSIS AND THE USE OF ENERGY BY CELLS (Pages 79–94)
Biological Order Is Made Possible by the Release of Heat Energy from Cells (Pages 79–82)

3–1 Easy, multiple choice

Living organisms require a continual supply of energy to exist because:

- A. they are defying the laws of thermodynamics.
- B. they convert it into heat energy which powers biosynthetic reactions.
- C. they are creating order out of disorder inside their cells.
- D. they are causing the entropy in the universe to decrease.
- E. they are closed systems isolated from the rest of the universe.

3–2 Easy, multiple choice (Requires information from section on pages 85–86)

Life is thermodynamically possible because living things:

- A. release heat to the environment.
- B. increase the degree of order in the universe.
- C. reproduce themselves.
- D. carry out energetically favorable reactions only.
- E. can carry out a chain of reactions that is energetically unfavorable.

3–3 Easy, multiple choice

The energy required by a human cell to grow and reproduce is provided by:

- A. the generation of order inside it.
- B. its anabolic metabolism.
- C. its catabolic metabolism.
- D. generation of heat.
- E. its biosynthetic reactions.

3–4 Intermediate, multiple choice (Requires information from sections on pages 79–94)

Which of the following statements are true?

- A. Resting cells do not produce any heat.
- B. Growing cells release less heat to the environment than resting cells because they use more energy.

C. The only processes that can occur to any significant degree are those that decrease the disorder of the universe.

D. Life is a thermodynamically spontaneous process.

E. Enzymes that couple unfavorable reactions to favorable reactions cause a decrease in total entropy.

Photosynthetic Organisms Use Sunlight to Synthesize Organic Molecules (Pages 82–83)

3–5   Easy, multiple choice

Which of the following statements about photosynthesis are true?

A. Photosynthesis is irrelevant to the existence of animals.

B. Because they can photosynthesize, plants require only CO_2, water and light to live.

C. The synthesis of sugars from CO_2 stops as soon as light is removed.

D. Photosynthesis generates the reducing agent NADPH.

E. Photosynthesis can be inhibited if the ratio of CO_2 to O_2 is very high.

Cells Obtain Energy by the Oxidation of Organic Molecules (Pages 83–84)

3–6   Intermediate, multiple choice

Which one or more of the following would contribute to a decrease in the amount of CO_2 in the atmosphere?

A. Turning rain forests into newspapers.

B. A huge flying saucer that completely eclipsed the sun.

C. Burning an *Essential Cell Biology* textbook.

D. Application of a fungicide to your garden soil.

E. Illuminating your pot plant with candles at night.

Oxidation and Reduction Involve Electron Transfers (Pages 84–85)

3–7 **Easy, short answer**

For each of the pairs A–D in Figure Q3–7, tick the more reduced member of the pair.

(i)

(ii)

(A) Fe^{3+}

Fe^{2+}

(B) $H_2C{=}CH_2$

$H_3C{-}CH_3$

(C)

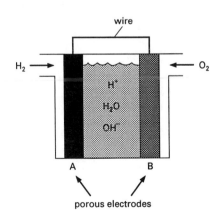

Q3–7

(D) $H{-}O{-}O{-}H$

$H{-}O{-}H$

3–8 **Difficult, short answer**

Some of the energy of oxidation-reduction reactions can be captured by separating the reactants and allowing electron transfer to occur through a wire. For the fuel cell diagrammed in Figure Q3–8, which one or more of the following events will happen?

A. Electrons will flow from electrode A to electrode B.

B. Electrons will flow from electrode B to electrode A.

C. The fuel cell will generate water.

D. The fuel cell will split water into hydrogen and oxygen.

Q3–8

Enzymes Lower the Barriers That Block Chemical Reactions (Pages 85–86)

3–9   Easy, multiple choice

The energy input required to initiate an energetically favorable chemical reaction is called the:

    A. free energy.
    B. activation energy.
    C. chemical bond energy.
    D. kinetic energy.
    E. potential energy.

3–10   Easy, multiple choice

Which of the following reactions are energetically favorable?

    A. base + sugar + phosphate → nucleotide
    B. amino acid + amino acid → peptide
    C. $CO_2 + H_2O$ → sugar
    D. sucrose → $CO_2 + H_2O$
    E. $N_2 + H_2$ → ammonia

3–11   Easy, data interpretation

An enzyme can increase the rate of conversion of substrate (S) to product (P) for the reaction depicted in Figure Q3–11 by:

    A. decreasing a.
    B. decreasing b.
    C. decreasing c.
    D. decreasing a and b by the
        same amount.
    E. decreasing b and c by the
        same amount.

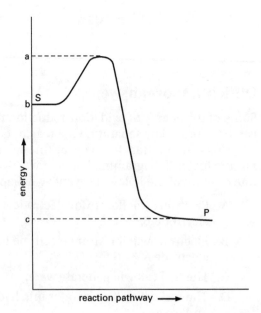

Q3–11

3–12   Easy, multiple choice

Which of the following statements about enzymes are correct?

A. Catalysis of an energetically unfavorable reaction by an enzyme will enable that reaction to occur.

B. An enzyme can direct a molecule along a particular reaction pathway.

C. An enzyme can catalyze many chemically different reactions.

D. An enzyme can bind to many structurally unrelated substrates.

E. Enzymes are permanently altered after catalyzing a reaction.

How Enzymes Find Their Substrates: The Importance of Rapid Diffusion (Pages 86–88)

3–13   Intermediate, data interpretation + short answer

(A) You are measuring the effect of temperature on the rate of an enzyme-catalyzed reaction. If you plot reaction rate against temperature, which of the graphs in Figure Q3–13 would you expect your plot to resemble?

(B) Explain why temperature has this effect.

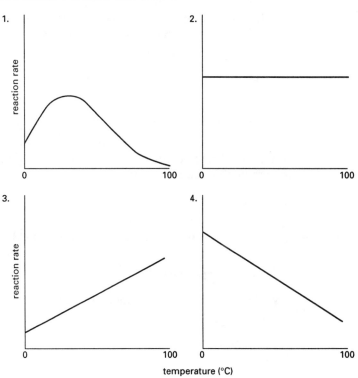

Q3–13

3–14   Intermediate, multiple choice (Requires information from section on pages 95–96, Chapter 1, and Panel 3–1)

Which of the following CANNOT be a reason that cells use enzymes rather than heat to increase the rate of biochemical reactions?

A. The temperature increase required to speed up a reaction by an appreciable extent is often huge.

B. Reactions cannot be coupled without enzymes.

C. An enzyme catalyzes just one or a very small number of different reactions; heat would affect all the reactions in a cell.

D. No organism can live at a temperature higher than 40°C.

E. Enzymes can accelerate reactions to a much greater extent than can heat.

The Free-Energy Change for a Reaction Determines Whether It Can Occur (Page 89)

3–15   Easy, multiple choice

Energetically favorable reactions are those that:

    A. decrease the entropy of a system.

    B. increase the free energy of a system.

    C. have a positive ΔG.

    D. decrease the free energy of a system.

    E. create order in a system.

3–16   Easy, short answer

(A) Which one or more of the following reactions will NOT occur spontaneously under the standard conditions that specify $\Delta G°$?

    1. ADP + P_i → ATP $\Delta G° = +7.3$ kcal/mole

    2. glucose-1-phosphate → glucose-6-phosphate $\Delta G° = -1.7$ kcal/mole

    3. glucose + fructose → sucrose $\Delta G° = +5.5$ kcal/mole

    4. glucose→ CO_2 + H_2O $\Delta G° = -686$ kcal/mole

(B) Which of the reactions in A could be coupled to any of the energetically un-favorable reactions to enable them to occur?

The Concentration of Reactants Influences ΔG (Pages 89–93)

3–17   Easy, multiple choice

Any reaction A ⇔ B is at equilibrium when:

    A. $\Delta G = 0$.

    B. $\Delta G° = 0$.

    C. [A] = [B].

    D. $\Delta G = \Delta G°$.

    E. both forward and backward rates reach zero.

3–18   Intermediate, multiple choice

The reaction X → Y has a small positive $\Delta G°$ at 37°C but can occur spontaneously in a human cell because:

    A. it is catalyzed by an enzyme.

    B. Y is rapidly used up by other reactions.

    C. the ratio [X]:[Y] is kept low in the cell.

    D. the thermal motion of molecules in the cell provides sufficient energy to overcome the activation energy barrier.

    E. the reaction generates heat.

3–19   Difficult, multiple choice

When the polymer X-X-X... is broken down into monomers, it is "phosphorylyzed" rather than hydrolyzed, in the reaction:

X-X-X... + P → X-P + X-X...   (reaction 1)

Given the $\Delta G°$ values of the reactions shown in Table Q3–19, what is the expected ratio of X-phosphate (X-P) to free phosphate (P) at equilibrium for reaction 1?

A. $1:10^6$.

B. $1:10^4$.

C. 1:1.

D. $10^4:1$.

E. $10^6:1$.

Table Q3–19

X-X-X... + H_2O → X + X-X...	$\Delta G° = -4.5$ kcal/mole
X + ATP → X-P + ADP	$\Delta G° = -2.8$ kcal/mole
ATP + H_2O → ADP + P	$\Delta G° = -7.3$ kcal/mole

For Sequential Reactions, $\Delta G°$ Values Are Additive (Pages 93–94)

3–20   Easy, multiple choice

The following reactions take place in a cell located next to a blood vessel.

X → Y $\Delta G° = -10$ kcal/mole

Y + O_2 → Z + CO_2 $\Delta G° = +0.5$ kcal/mole

Normally, the blood vessel brings in oxygen and takes away carbon dioxide, but years of overindulgence have taken their toll, and it has become completely clogged with cholesterol, cutting off the blood supply. Which of the following molecules would be expected to accumulate in large amounts?

A. X.

B. Y.

C. Z.

D. Y and Z.

E. X and Z.

ACTIVATED CARRIER MOLECULES AND BIOSYNTHESIS (Pages 95–104)
The Formation of an Activated Carrier Is Coupled to an Energetically Favorable Reaction (Pages 95–96)

3–21   Intermediate, short answer

You are studying a metabolic pathway in which a substrate X is oxidized with an overall ΔG of –600 kcal/mole. You find that for the oxidation of 1 mole of substrate X, 10 moles of ATP are synthesized from ADP + P_i (ΔG for the ATP synthesis reaction in your conditions is +12 kcal/mole) and 280 kcal of energy are released as heat. What is the most likely explanation for the failure to account for all of the energy in your results?

ATP Is the Most Widely Used Activated Carrier Molecule (Pages 96–97)

3–22   Easy, multiple choice

A common means of providing energy to an energetically unfavorable reaction in a cell is by:

A. generation of a higher temperature by the cell.
B. transfer of a phosphate group from the substrate to ADP.
C. enzyme catalysis of the reaction.
D. coupling of ATP hydrolysis to the reaction.
E. coupling of the synthesis of ATP to the reaction.

Energy Stored in ATP Is Often Harnessed to Join Two Molecules Together (Pages 97–98)

3–23   Intermediate, multiple choice

An anhydride formed between a carboxylic acid and a phosphate (Figure Q3–23A) is formed as a high-energy intermediate in some reactions in which ATP is used as the energy source. Arsenate mimics phosphate and can also be incorporated into a similar high-energy intermediate (Figure Q3–23B). The reaction profiles for the hydrolysis of these two high-energy intermediates are given in Figure Q3–23C. What is the effect of substituting arsenate for phosphate in this reaction?

A. It forms a high-energy intermediate of lower energy.
B. It forms a high-energy intermediate of the same energy.
C. It decreases the stability of the high-energy intermediate.
D. It increases the stability of the high-energy intermediate.
E. It has no effect on the stability of the high-energy intermediate.

Q3–23   (A)   (B)   (C)

3–24   Intermediate, short answer

Glutamine synthesis occurs through a phosphorylated intermediate as shown in Figure Q3–24. This intermediate is formed in step 1 by the transfer of a phosphate from ATP, leaving ADP as a product. Is it therefore necessary that the bond to phosphate in the phosphorylated intermediate be of lower energy than the phosphoanhydride bond in ATP?

Q3–24

glutamic acid (glutamate)	high-energy intermediate	glutamine

3–25   Difficult, multiple choice

You are studying a biochemical pathway that requires ATP as an energy source. To your dismay, the reactions soon stop, partly because the ATP is rapidly used up and partly because an excess of ADP builds up and inhibits the enzymes involved. You are about to give up when the following table from a biochemistry textbook catches your eye.

Hydrolysis reaction	ΔG°
enzyme A creatine + ATP → creatine phosphate + ADP	+3 kcal/mole
enzyme B ATP + H_2O → ADP + phosphate	–7.3 kcal/mole
enzyme D pyrophosphate + H_2O → 2 phosphate	–7 kcal/mole
enzyme E glucose 6-phosphate + H_2O → glucose + phosphate	–3.3 kcal/mole

Which of the following reagents are most likely to revitalize your reaction?

    A. A vast excess of ATP.

    B. Glucose 6-phosphate and enzyme E.

    C. Creatine phosphate and enzyme A.

    D. Pyrophosphate.

    E. Pyrophosphate and enzyme D.

NADH and NADPH Are Important Electron Carriers (Pages 98–100)

3–26   Intermediate, multiple choice

Which of the following statements are true?

A. The oxidation of food molecules generates NAD^+.

B. NADH and NADPH are found in mutually exclusive parts of the cell.

C. The ratio of $NADPH:NADP^+$ is higher than the ratio of $NADH:NAD^+$ because each molecule of NADPH is a stronger reducing agent than a molecule of NADH.

D. Many enzymes can use NADPH and NADH interchangeably.

E. One molecule of NADPH can cause the transfer of two hydrogen atoms.

There Are Many Other Activated Carrier Molecules in Cells (Pages 100–103)

3–27   Intermediate, matching/fill in blanks

Match the activated carrier molecules in List A with the groups they transfer, selected from List B. Write the appropriate number beside each item in List A.

List A	List B
A. ATP.	1. $-COO^-$.
B. Acetyl CoA.	2. e^- and H^+.
C. NADPH.	3. Glucose.
D. Activated biotin.	4. $-PO_4^{3-}$.
E. *S*-adenosylmethionine.	5. $-CH_3$.
	6. Nucleotide.
	7. $-COCH_3$.
	8. Amino acid.

The Synthesis of Biological Polymers Requires an Energy Input (Pages 103–104)

3–28   Easy, multiple choice (Requires information from Chapter 2)

Which of the following processes must be coupled to an energetically favorable reaction in order to occur?

A. Conversion of protein into amino acids.

B. Polymerization of amino acids into polypeptides.

C. Conversion of glucose to carbon dioxide and water.

D. Formation of a bilayer from phospholipids in water.

E. The hydrolysis of ATP.

3–29 **Intermediate, multiple choice**

The enzymes that catalyze the synthesis of macromolecules do not also catalyze their break-down by hydrolysis because:

    A. enzymes can catalyze reactions in only one direction.

    B. hydrolysis is not an energetically favorable reaction.

    C. the hydrolytic reaction is not the reverse of the reaction pathway that is used for biosynthesis.

    D. enzymes are destroyed immediately after synthesis is completed.

    E. biosynthesis proceeds more rapidly than hydrolysis.

3–30 **Intermediate, short answer**

The energy required for the addition of a C nucleotide subunit (CMP) to a growing poly-nucleotide chain is originally derived from the hydrolysis of ATP. Explain how this is achieved.

Answers

A3–1.   C. A is untrue as no system, living or otherwise, can defy the laws of thermodynamics. B is untrue as living organisms do not use heat to power biochemical reactions. Heat is produced in the course of biochemical reactions. D is untrue: although living organisms are causing a local decrease in entropy, they cannot cause a decrease in the entropy of the universe as a whole as that would be a thermodynamic impossibility. E is untrue as living organisms are not closed systems.

A3–2.   A. By releasing heat to their environment, living things increase the entropy of the environment, so compensating for the decrease in entropy inside cells. They thus satisfy the second law of thermodynamics.

A3–3.   C. Catabolic reactions are the reactions in which a cell breaks down food molecules, releasing the energy held within their chemical bonds. A, B, and E are energy-requiring processes.

A3–4.   D. Life is a thermodynamically spontaneous process, or else it would not exist. The increase in entropy of the surroundings that accompanies the chemical reactions that go into living creatures makes life thermodynamically spontaneous. A is false because all living cells produce heat, regardless of whether they are growing or not. B is false because the amount of order created by the synthesis of molecules required for growth must be compensated for by an increase in the heat released to the environment; growing cells do require more energy (which they obtain by consuming more food), but much of this extra energy is released as heat. C states the opposite of the second law of thermodynamics, and is thus false. No enzyme can cause a decrease in the entropy of the universe, making E false.

A3–5.   D. A is false, since animals depend ultimately on the photosynthetic organisms at the bottom of food chains to produce a continuing source of food. B is false, since plants also need sources of nitrogen and other elements. C is false: the synthesis of sugars from carbon dioxide can take place in the absence of light, as long as there is a source of energy. E is false: since CO_2 is one of the inputs to photosynthesis and O_2 is one of the products, photosynthesis would be inhibited if the ratio of O_2 to CO_2 were high, not as stated.

A3–6.   D. Fungi are not photosynthetic and thus carry out respiration only, oxidizing food molecules and releasing CO_2. A and B will decrease the amount of photosynthesis and thus increase the amount of CO_2 available, while C will release CO_2 by oxidation of organic material. D will cause the plant to photosynthesize and take in CO_2, but burning the candles will also release CO_2, and since very little of the chemical bond energy stored in wax will be released as light and only a fraction of this will be absorbed by the plants and used to drive photosynthesis, you will end up generating more CO_2 by combustion than your plant will consume during photosynthesis.

A3–7.   A, ii; B, ii; C, i; D, ii. "More reduced" means having more electrons; gain of electrons can result in an increased negative charge, a decreased positive charge and can be due to an increase in the number of hydrogen atoms in a molecule.

A3–8.   A and C. H_2 tends to donate electrons to O_2, so electrons will tend to flow from A to B. The net reaction driving the fuel cell is $2H_2 + O_2 \rightarrow 2H_2O$.

A3–9.   B.

A3–10. D.

A3–11.   A. Enzymes can affect only the activation energy (a–b) of a process.

A3–12.   B.

A3–13.   (A) Graph 1. (B) By increasing thermal motion, increasing the temperature increases the rate of diffusion of components and the number of collisions of sufficient energy to overcome the activation energy. An increase in temperature will thus increase the reaction rate initially. However, enzymes are proteins and are held together by noncovalent interactions, so at very high temperatures, the enzyme will begin to denature and the reaction rate will fall.

A3–14.   D. There are bacteria that live in hot springs at temperatures of more than 80°C. All the other statements are true.

A3–15.   D.

A3–16.   (A) 1 and 3. Only reactions with a negative ΔG can occur spontaneously. (B) Coupling of reaction 4 to either of the reactions 1 or 3 would provide an overall negative ΔG for the coupled reactions, thus enabling them to occur.

A3–17.   A. The value of ΔG for the reaction A $\Leftrightarrow$ B is zero when there is no *net* tendency for either A $\rightarrow$ B or B $\rightarrow$ A, which is the definition of equilibrium. $\Delta G°$ is a constant and is thus always the same regardless of whether the reaction has reached equilibrium or not. Thus B and D are incorrect. C is an incorrect answer; although a particular reaction might be at equilibrium when the concentration of substrate equalled that of product, this is not true for most reactions. E is not a definition of equilibrium, but of a reaction that is not occurring at all.

A3–18.   B. The value of ΔG (but not $\Delta G°$) is influenced by the concentration of the reactants. Thus in cellular conditions, a reaction with a small positive $\Delta G°$ might have a small negative ΔG if the concentration of product is kept very low compared with the concentration of substrate. A is irrelevant; enzyme catalysis cannot by itself enable an energetically unfavorable reaction to occur. C states the opposite condition to B and thus would not enable the reaction X $\rightarrow$ Y to occur; in such conditions the reaction Y $\rightarrow$ X might be more likely. D would not be true. E is irrelevant.

A3–19.   C. Reaction 1 can be written as the sum of the three reactions given, since the ATP used in Step 2 is restored in Step 3.

$$X\text{-}X\text{-}X\ldots + H_2O \rightarrow X + X\text{-}X\ldots \qquad \Delta G° = -4.5 \text{ kcal/mole}$$
$$X + ATP \rightarrow X\text{-}P + ADP \qquad \Delta G° = -2.8 \text{ kcal/mole}$$
$$ADP + P \rightarrow ATP + H_2O \qquad \Delta G° = +7.3 \text{ kcal/mole}$$

Since $\Delta G°$ values are additive, $\Delta G°^{total} = 0$, and if $\Delta G° = 0$, $K_{eq} = 1$, meaning that [products]/[reactants] = 1, and the ratio of X-P to P is 1:1.

A3–20.   B. The constant removal of CO_2 and replenishment of O_2 by the blood normally drives the reaction Y $\rightarrow$ Z. Therefore, when CO_2 is allowed to accumulate and O_2 drops, Y will accumulate. Because the $\Delta G°$ of the first reaction is very negative, Y can accumulate to a very high level without causing significant amounts of X to build up.

A3–21.   The synthesis of 10 moles of ATP will use 120 kcal of the 600 kcal of usable energy released by the oxidation of 1 mole of substrate X. Taking away the energy released as heat still leaves 200 kcal of free energy unaccounted for. The most likely explanation for this discrepancy is that energy-requiring reactions other than ATP synthesis are also directly coupled to the oxidation of substrate X.

A3–22.   D.

A3–23.   C. The activation energy of the arsenate compound is extremely low, as can be seen from the reaction profile, meaning that its high-energy intermediate is very unstable and will be spontaneously hydrolyzed more rapidly than the phosphate compound. In fact, this hydrolysis occurs rapidly without enzyme catalysis, even in cellular conditions. Thus D and E are untrue. A and B are untrue as more energy is released by the hydrolysis of the arsenoanhydride bond (as inferred by the greater difference in energy level between reactants and products in Figure Q3–23) so, by definition, the arsenoanhydride bond is said to have more energy than the phosphoanhydride bond.

A3–24.   No. In fact, the anhydride bond to carbon in the intermediate is of a higher energy. But the reaction is driven to produce glutamine by the large release of free energy when ammonia is added in Step 2.

A3–25.   C. An excess of ATP will initially restore the reactions, but as ATP is hydrolyzed, ADP will build up and inhibit the enzymes again. Pyrophosphate does not look like ATP and is therefore unlikely to be used by the enzymes as an alternative energy source. Pyrophosphate + enzyme D will just heat things up. What you need is a high-energy source of phosphate that can convert ADP back to ATP. Since the $\Delta G°$ of the reaction,

$$\text{ATP} + \text{creatine} \rightarrow \text{ADP} + \text{creatine phosphate},$$

catalyzed by enzyme A is greater than zero, the addition of creatine phosphate and enzyme A can be used to form ATP from ADP, regenerating the ATP while also forming creatine as a waste product.

A3–26.   E. NADPH has two electrons and one proton more than $NADP^+$ and donates both electrons. Protons are always present in solution. So the recipient molecule effectively acquires two hydrogen atoms. A is false because oxidation of food molecules produces NADH, not NAD^+. B is false: NADH and NADPH can be found in the same parts of the cell, but are used for different functions; this is possible because the enzymes that recognize one do not recognize the other; thus D is also false. C is false: the parts of NADPH and NADH that participate in reduction are identical, and thus they both have essentially the same reducing power.

A3–27.   A, 4 (phosphate group); B, 7 (acetyl group); C, 2; D, 1 (carboxyl group); E, 5 (methyl group).

A3–28.   B. Polymerization of amino acids into a polypeptide leads to the formation of peptide bonds that have higher energy than the free amino acids and also represents an increase in order. Hence, it can only be brought about via an input of energy. The other processes represent a decrease in order and are thermodynamically spontaneous.

A3–29.   C. Hydrolysis is not the reverse of the reactions catalyzed by biosynthetic enzymes. For instance, the reactions involved in RNA biosynthesis are:

$$\text{polynucleotide}_{(n)} + \text{NTP} \rightarrow \text{polynucleotide}_{(n+1)} + \text{PP}_i$$

$$\text{PP}_i + \text{H}_2\text{O} \rightarrow 2\,\text{P}_i.$$

The reverse reactions are:

$$2\text{P}_i \rightarrow \text{PP}_i + \text{H}_2\text{O}$$

$$\text{PP}_i + \text{polynucleotide}_{(n+1)} \rightarrow \text{polynucleotide}_{(n)} + \text{NTP},$$

not

$$\text{polynucleotide}_{(n+1)} + H_2O \rightarrow \text{polynucleotide}_{(n)} + NMP \text{ (nucleoside monophosphate)},$$

which is the reaction by which RNA is hydrolyzed. A and B are untrue: enzymes catalyze both forward and reverse reactions, and hydrolysis is an energetically favorable reaction. D is untrue, since enzymes are unchanged by participating in catalysis. Whether E is true or not for any particular reaction is irrelevant.

A3–30.   The free C nucleotide before it is added to the polynucleotide chain is in the form of CTP (cytidine triphosphate), which has been synthesized from CMP by the sequential transfer of two terminal phosphate groups from two molecules of ATP. This is where the ATP hydrolysis comes in. The reaction that adds guanine monophosphate to the polynucleotide chain releases pyrophosphate (PP_i), which is then hydrolyzed to inorganic phosphate, thus making available sufficient energy to drive the overall condensation reaction.

4 How Cells Obtain Energy from Food

Questions

THE BREAKDOWN OF SUGARS AND FATS (Pages 108–110)
Food Molecules Are Broken Down in Three Stages to Produce ATP (Pages 110–112)

4–1 Easy, multiple choice

Which of the following stages in the breakdown of the piece of toast you had for breakfast generates the most ATP?

    A. Digestion of starch to glucose.

    B. Glycolysis.

    C. The citric acid cycle.

    D. Oxidative phosphorylation.

    E. Conversion of pyruvate to acetyl CoA.

4–2 Easy, multiple choice

The advantage to the cell of the gradual oxidation of glucose during cellular respiration compared with its combustion to CO_2 and H_2O in a single step is:

    A. more free energy is released for a given amount of glucose oxidized.

    B. no energy is lost as heat.

    C. energy can be extracted in usable amounts.

    D. more CO_2 is produced for a given amount of glucose oxidized.

    E. less O_2 is required for a given amount of glucose oxidized.

4–3 Easy, multiple choice

The final metabolite produced by glycolysis is:

    A. acetyl CoA.

    B. pyruvate.

    C. 3-phosphoglycerate.

    D. glyceraldehyde 3-phosphate.

    E. fatty acids.

4–4   Easy, multiple choice (Requires information from sections on pages 120–121 and pages 127–129)

In aerobic respiration, carbon dioxide is produced during:

    A. breakdown of glycogen.

    B. glycolysis.

    C. conversion of pyruvate to acetyl CoA.

    D. oxidative phosphorylation.

    E. the citric acid cycle.

4–5   Intermediate, multiple choice

On a diet consisting of nothing but protein, which of the following is the most likely outcome?

    A. Loss of weight because amino acids cannot be used for the synthesis of fat.

    B. Muscle gain because the amino acids will go directly into building muscle.

    C. Tiredness because amino acids cannot be used to generate energy.

    D. Excretion of more nitrogenous (ammonia-derived) wastes than with a more balanced diet.

    E. Production of more carbon dioxide than with a more balanced diet.

4–6   Intermediate, art labeling (Requires student to have studied the whole chapter and also Chapter 1)

(A) Figure Q4–6 represents a cell lining the gut. Draw numbered labeled lines to locate exactly where the following processes should take place.

1. Glycolysis.

2. Citric acid cycle.

3. Conversion of pyruvate to acetyl groups.

4. Oxidation of fatty acids to acetyl CoA.

5. Synthesis of fats.

6. Glycogen breakdown.

7. Glycogen synthesis.

8. Protein digestion.

(B) Indicate precisely where oxidative phosphorylation takes place.

(C) In a plant cell, which of the reactions listed above might take place in chloroplasts.

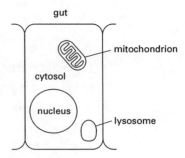

Q4–6

Glycolysis Is a Central ATP-producing Pathway (Pages 110–114)

4–7   Easy, multiple choice

The oxidation of sugars by glycolysis:

    A. occurs only in aerobic organisms.

    B. generates carbon dioxide.

    C. produces a net gain of ATP.

    D. occurs in mitochondria.

    E. uses NADH as a source of energy.

4–8   Intermediate, multiple choice (Requires information from Panel 4–1)

What purpose is served by the phosphorylation of glucose to glucose 6-phosphate by the enzyme hexokinase as the first step in glycolysis?

    A. It helps cells keep glucose in the cytoplasm.

    B. It generates a high-energy phosphate bond.

    C. It converts ATP to a more useful form.

    D. It enables the glucose 6-phosphate to be recognized by phosphofructokinase, the next enzyme in the glycolytic pathway.

4–9   Intermediate, multiple choice (Requires information from Panel 4–1)

Which reaction does the enzyme phosphoglucose isomerase catalyze?

    A. glucose → glucose 6-phosphate.

    B. fructose 6-phosphate → fructose 1,6-bisphosphate.

    C. glucose 6-phosphate → fructose 6-phosphate.

    D. glucose → glucose 1-phosphate.

    E. glucose → fructose.

4–10   Intermediate, multiple choice (Requires information from Panel 4–1)

Which of the following enzymes control the entry of sugars into the glycolytic pathway?

    A. Hexokinase.

    B. Phosphofructokinase.

    C. Pyruvate dehydrogenase.

    D. Isocitrate dehydrogenase.

    E. Phosphoglucose isomerase.

4–11   Intermediate, art labeling (Requires information from Panel 4–1)

Give the full names of the reactants indicated by question marks in Figure Q4–11.

Q4–11

Fermentations Allow ATP to Be Produced in the Absence of Oxygen (Page 114)

4–12   Easy, multiple choice (Requires information from Chapter 1)

Which of the following cells rely exclusively on glycolysis to supply them with ATP?

    A. Anaerobically growing yeast.
    B. Aerobic bacteria.
    C. Skeletal muscle cells.
    D. Plant cells.
    E. Protozoa.

4–13   Easy, multiple choice

In anaerobic conditions, skeletal muscle produces:

    A. lactate and CO_2.
    B. ethanol and CO_2.
    C. lactate only.
    D. ethanol only.
    E. lactate, ethanol, and CO_2.

4–14   Intermediate, multiple choice

In mammals, liver cells are able to convert lactate to pyruvate. What purpose does this serve for the organism?

    A. It is an important way of generating NADH.
    B. It is an important way of generating NAD^+.
    C. It allows the organism to grow in anaerobic conditions.
    D. It allows more energy to be extracted from fermented carbon compounds.
    E. It is an important way for the body to generate heat.

4–15   Easy, short answer

Anaerobically growing yeast further metabolize the pyruvate produced by glycolysis to CO_2 and ethanol in a series of fermentation reactions.

(A) What other important reaction occurs during this fermentation?

(B) Why is this reaction (i.e., the answer to part A) essential for the anaerobically growing cell?

Glycolysis Illustrates How Enzymes Couple Oxidation to Energy Storage (Pages 114–118)

4–16   Difficult, multiple choice (Requires information from Panel 4–1 and Figure 4–5)

In the first energy-generating steps in glycolysis, glyceraldehyde 3-phosphate undergoes an energetically favorable reaction in which it is simultaneously oxidized and phosphorylated by the enzyme glyceraldehyde 3-phosphate dehydrogenase to form 1,3-bisphosphoglycerate, with the accompanying conversion of NAD^+ to NADH. In a second energetically favorable reaction, catalyzed by a second enzyme, 1,3-bisphosphoglycerate is then converted to 3-phosphoglycerate, with the accompanying conversion of ADP to ATP. Which of the following statements are true?

  A. The reaction glyceraldehyde 3-phosphate → 1,3-bisphosphoglycerate should be inhibited when levels of NADH fall.

  B. The $\Delta G°$ for the oxidation of the aldehyde group on glyceraldehyde 3-phosphate to form a carboxylic acid is more negative than the $\Delta G°$ for ATP hydrolysis.

  C. The high-energy bond to the phosphate group in glyceraldehyde 3-phosphate contributes to driving the reaction forward.

  D. The cysteine side chain on the enzyme is oxidized by NAD^+.

  E. The overall reaction glyceraldehyde 3-phosphate → 3-phosphoglycerate has a positive ΔG.

4–17   Intermediate, multiple choice

The simultaneous oxidation and phosphorylation of glyceraldehyde 3-phosphate described in Question 4–16 involves the formation of a highly reactive covalent thioester bond between a cysteine side chain (reactive group –SH) on the enzyme (glyceraldehyde 3-phosphate dehydrogenase) and the oxidized intermediate, as shown arrowed in Figure Q4–17A. If the enzyme had a serine (reactive group –OH) instead of a cysteine at this position, which could form only a much lower-energy bond to the oxidized substrate, as shown arrowed in Figure Q4–17B, how might this new enzyme act?

  A. It would oxidize the substrate and phosphorylate it without releasing it.

  B. It would oxidize the substrate but not release it.

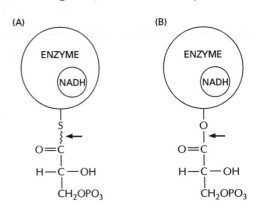

Q4–17

C. It would phosphorylate the substrate on the 2 position instead of the 1 position.

D. It would behave just like the normal enzyme.

Sugars and Fats Are Both Degraded to Acetyl CoA in Mitochondria (Pages 118–119)

4–18   Easy, multiple choice

Acetyl CoA is:

A. synthesized from pyruvate in the mitochondrial intermembrane space.

B. the intermediate through which food molecules are completely metabolized to carbon dioxide in animal cells.

C. synthesized from pyruvate and CoA in a reaction that also generates NADH, CO_2 and ATP.

D. synthesized by the breakdown of fatty acids in the cytosol.

E. an intermediate in the oxidation of glucose in anaerobic skeletal muscle.

4–19   Intermediate/difficult, short answer (Requires information from section on pages 126–127)

Assuming complete oxidation, which of the fatty acids shown in Figure Q4–19 will generate the most ATP through cellular respiration?

(A) $CH_3-CH_2-CH_2-CH=CH-C\overset{O}{\underset{OH}{}}$

(B) $CH_3-CH_2-CH_2-CH_2-CH_2-C\overset{O}{\underset{OH}{}}$

Q4–19

The Citric Acid Cycle Generates NADH by Oxidizing Acetyl Groups to CO_2 (Pages 119–124)

4–20   Easy, multiple choice

During each turn of the citric acid cycle:

A. the two carbon atoms of the acetyl CoA that enters the cycle are completely oxidized to CO_2.

B. three molecules of ATP are generated.

C. three molecules of NADH are generated.

D. an acetyl group is added to citric acid.

E. three molecules of CO_2 are generated.

4–21 Intermediate, short answer

Explain why the following statement is untrue.

"One mole of oxaloacetate is required for every mole of acetyl CoA that is metabolized via the citric acid cycle."

4–22 Easy, multiple choice

Cells oxidizing acetyl groups via the citric acid cycle require molecular oxygen in order to:

    A. oxidize the acetyl groups to CO_2.

    B. regenerate NAD^+.

    C. regenerate $FADH_2$.

    D. regenerate CoA.

    E. oxidize fatty acids to acetyl groups.

4–23 Difficult, short answer (Requires information from Panel 4–2)

Given a mixture of all the enzymes of the citric acid cycle plus acetyl CoA, which of the following sets of additions could support conversion of acetyl CoA to carbon dioxide? Explain why.

    A. Water, NAD^+, GDP, phosphate, FAD^+.

    B. Water, NAD^+, GDP, phosphate, FAD^+, oxaloacetate.

    C. Water, NAD^+, GDP, phosphate, FAD^+, citrate.

    D. Water, NAD^+, GDP, phosphate, FAD^+, citrate, coenzyme A.

4–24 Difficult, multiple choice

A deficiency of the vitamin thiamine causes elevated levels of both pyruvate and α-ketoglutarate in the blood. What is a likely role for thiamine in general metabolism?

    A. It is required for the use of NAD^+ as a substrate in oxidation-reduction reactions.

    B. It is required for the function of certain dehydrogenases.

    C. It is required for the synthesis of coenzyme A.

    D. It is an inhibitor of isocitrate dehydrogenase.

Electron Transport Drives the Synthesis of the Majority of the ATP in Most Cells (Pages 124–125)

4–25 Easy, multiple choice

Which of the following statements regarding electron transport are true?

    A. Only high-energy electrons from NADH can be used to drive the electron transport chain.

B. The proteins involved in electron transport couple oxidation to phosphorylation in much the same way that glyceraldehyde 3-phosphate dehydrogenase couples oxidation and phosphorylation in glycolysis.

C. Electron transport occurs only in eucaryotes.

D. Molecular oxygen is required in order to donate electrons to the electron transport chain.

E. Electrons passing along the electron transport chain move to successively lower energy states.

4–26   Intermediate, multiple choice

In the final stage of the oxidation of food molecules, a gradient of protons is formed across the inner mitochondrial membrane, which is normally impermeable to protons. If cells are exposed to an agent that causes the membrane to become freely permeable to protons, which of the following effects would you expect to observe?

A. Cells would be completely unable to synthesize ATP.

B. NADH would build up.

C. Carbon dioxide production would cease.

D. Consumption of oxygen would fall.

E. The ratio of ATP to ADP in the cytoplasm would fall.

STORING AND UTILIZING FOOD (Pages 125–129)
Organisms Store Food Molecules in Special Reservoirs (Pages 125–127)

4–27   Easy, multiple choice

Which of the following statements are true?

A. Plant cells store all their food reserves as starch, whereas animals store all their food reserves as glycogen.

B. Glycogen stores more energy than starch because glycogen molecules have many more branch points that can be hydrolyzed.

C. Animal cells can convert fatty acids to sugars.

D. Plants synthesize starch for the same reason that animals synthesize glycogen.

E. Protein is an important form of energy storage in animal cells under normal conditions.

4–28   Easy, multiple choice

Glycogen is a more useful food storage molecule than fat because:

A. a gram of glycogen produces more energy than a gram of fat.

B. it can be utilized under anaerobic conditions whereas fat cannot.

C. it binds water and therefore is useful in keeping the body hydrated.

D. for the same amount of energy storage, glycogen occupies less space in a cell than does fat.

E. glycogen can be carried to cells via the bloodstream whereas fats cannot.

Many Biosynthetic Pathways Begin with Glycolysis or the Citric Acid Cycle (Pages 127–128)

4–29   Intermediate, multiple choice

The intermediates of the citric acid cycle are constantly being depleted because they are used to produce many of the amino acids needed to make proteins. These intermediates must therefore be replenished by the conversion of pyruvate to oxaloacetate by the enzyme pyruvate carboxylase. Bacteria, but not animal cells, have additional enzymes that can carry out the reaction acetyl CoA + isocitrate → oxaloacetate + succinate. Which of the following compounds will not support the growth of animal cells when used as the major source of carbon in the medium, but will support the growth of nonphotosynthetic bacteria?

A. Pyruvate.

B. Glucose.

C. Fatty acids.

D. Carbon dioxide.

E. Fructose.

4–30   Intermediate, short answer (Requires students to have studied the whole chapter)

A carbon atom in a carbon dioxide molecule in the atmosphere eventually becomes part of one of the enzymes that catalyze glycolysis in one of your cells. Some of the compounds it passed through on the way are listed below. Place these compounds in the order in which the carbon atom would have passed through them, given that it passed through each molecule only once. Indicate at which point the carbon atom entered your body.

$$CO_2 \rightarrow 1 \rightarrow 2 \rightarrow 3 \rightarrow 4 \rightarrow 5 \rightarrow 6 \rightarrow \text{enzyme}$$

amino acid; citrate; glycogen; starch; fructose 1,6-bisphosphate; pyruvate.

Answers

A4–1. D. Oxidative phosphorylation produces about 28 ATP molecules. A produces no ATP; B nets 2 ATP; C produces 1 GTP; and E produces no ATP.

A4–2. C. A is untrue as the same overall amount of free energy is released by glucose oxidation, whatever the route. B is untrue as a proportion of the energy released is still lost as heat. D and E are untrue as the same amount of CO_2 will be released and O_2 consumed by the oxidation of glucose to CO_2 and H_2O, whatever the route.

A4–3. B.

A4–4. C. CO_2 is also produced by the citric acid cycle.

A4–5. D. Because ammonia is given off when amino acids are metabolized to yield energy but is not given off when sugars and fats are metabolized, you would expect more nitrogenous waste to be excreted. C is incorrect since amino acids can be converted into pyruvate and acetyl CoA and used to generate energy. If more amino acids are consumed than are used, the body will not store them as protein in muscle tissue but rather will store them as fat, so A and B are incorrect. Amino acid metabolism does not produce more carbon dioxide than carbohydrate or fat metabolism.

A4–6. (A) Figure A4–6. (B) Labeling line for oxidative phosphorylation should point to inner mitochondrial membrane. (C) Synthesis of fats (5).

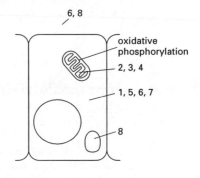

A4–6

A4–7. C.

A4–8. A. B is incorrect since the phosphate transferred to the glucose is not in a high-energy linkage and in this form is not used to provide energy in later steps. C is incorrect since in the reaction catalyzed by hexokinase, ATP is converted to ADP, which is not useful as an energy source for most cellular reactions, even though it still has one high-energy bond. D is incorrect since the next enzyme in the pathway is phosphoglucose isomerase, which converts the glucose 6-phosphate to fructose 6-phosphate, not phosphofructokinase.

A4–9. C. The isomerase part of the enzyme name indicates that it catalyzes an isomerization reaction, and the phosphoglucose part of the name indicates the type of substrate used. The enzyme that catalyzed reaction E would be called glucose isomerase.

A4–10. B. The phosphorylation of fructose 6-phosphate to fructose 1,6-bisphosphate, catalyzed by phosphofructokinase, is the first irreversible step unique to glycolysis and is the point of regulation.

A4–11. 1, phosphoenolpyruvate; 2, adenosine diphosphate; 3, pyruvate; 4, adenosine triphosphate.

A4–12. A. All the other cells can carry out oxidative phosphorylation.

A4–13. C.

A4–14. D. Lactate is a metabolic dead end and, unless it is converted to a molecule that can be further broken down, it cannot be utilized any further by cells; pyruvate, on the other hand, can be converted to acetyl CoA and metabolized in the citric acid cycle. Conversion of lactate to pyruvate uses up NAD^+ (making B incorrect) and generates NADH, but the reason animal cells make lactate in the first place is that they are suffering from an excess of NADH (and thus A is incorrect). C is incorrect since no mammal can survive without oxygen. The amount of heat generated by converting lactate to pyruvate is negligible, so E is incorrect.

A4–15. (A) NADH $\rightarrow$ NAD^+. (B) Under anaerobic conditions, it is the only means of generating the NAD^+ required for glycolysis, the main energy-generating pathway of an anaerobically growing yeast cell.

A4–16. B. This is another way of stating that the energetically favorable oxidation of glyceraldehyde 3-phosphate provides sufficient energy to ultimately drive the energy-requiring step of ATP synthesis from ADP. A is untrue: NADH is an end product of the reaction G-3-P to 1,3-PG and therefore high (not low) levels of it would inhibit the reaction. C is untrue: the reactions do not involve the 3-phosphate group on glyceraldehyde 3-phosphate at all. D is untrue, since the cysteine on the enzyme is important in making a covalent intermediate with the substrate and is not oxidized by NAD^+. E is untrue, since if the reaction had an overall positive ΔG, it could not be used to power the energetically unfavorable reactions of ATP and NADH synthesis.

A4–17. B. The phosphorylation and release of the product from the normal enzyme is possible because the high-energy thioester bond formed between the oxidized substrate and enzyme can be attacked by a phosphate molecule. If the bond between oxidized substrate and enzyme is a much lower-energy bond, the enzyme will not be able to transfer the oxidized substrate to a phosphate group, and substrate and enzyme will remain covalently bound. A, C, and D could not happen, as none of the bonds in the substrate molecule is reactive enough to be broken by a phosphate group.

A4–18. B. Some food molecules can enter the citric acid cycle at points other than acetyl CoA, but only molecules that enter as acetyl CoA are completely oxidized to carbon dioxide. Acetyl CoA is made in the mitochondrial matrix, not the intermembrane space, so A is incorrect. C is incorrect since no ATP is directly generated by the conversion of pyruvate to acetyl CoA; carbon dioxide and NADH are the only other products of the reaction. D is incorrect as the breakdown of fatty acids to produce acetyl CoA occurs in the mitochondrial matrix. E is incorrect since acetyl CoA is not an intermediate in the anaerobic fermentation reaction that coverts pyruvate to lactate.

A4–19. B. This will produce 2 NADH, 2 $FADH_2$, and 3 acetyl CoA on complete oxidation. Because of the double bond in A, this fatty acid will produce 2 NADH and 3 acetyl CoA, but only 1 $FADH_2$, since the initial oxidation step using FAD^+ that reduces the two-carbon unit $–CH_2CH_2–$ to C=C is not needed for two-carbon units already containing a double bond. So although the amount of acetyl CoA entering the citric acid cycle will be the same for A and B, fewer reducing equivalents will eventually enter the electron transport chain from the oxidation of A and thus less ATP will be produced.

A4–20. C.

A4–21. Only small amounts of oxaloacetate are required relative to the amount of acetyl CoA metabolized because oxaloacetate is regenerated after every round of the citric acid cycle.

A4–22.   B. The citric acid cycle generates high-energy electrons that are passed to NAD^+ to form NADH. NADH then donates these electrons to the electron transport chain that drives oxidative phosphorylation, regenerating NAD^+. The electrons from NADH are passed via the electron transport chain to oxygen.

A4–23.   B and D. To get the cycle turning you need water, NAD^+, GDP, phosphate, FAD, coenzyme A, and at least one intermediate of the citric acid cycle, which are all provided in D. In addition, B will be sufficient since oxaloacetate reacts with acetyl CoA to release a molecule of coenzyme A, which can then be reused. C would produce a small amount of CO_2 initially from the added citrate, but the cycle could not continue since citrate has to go through a step requiring coenzyme A to complete the cycle. A will not work, as this set does not include any citric acid cycle intermediate.

A4–24.   B. The immediate fate of both pyruvate and α-ketoglutarate is to be converted into other metabolites in similar reactions which are catalyzed by dehydrogenases. So you would expect them both to build up if thiamine were required for the action of these dehydrogenases. A and C are unlikely, since if thiamine were required for the synthesis or utilization of such widely used carrier molecules as NAD^+ and coenzyme A, many metabolic steps would be affected and you would see a buildup of more than just pyruvate and α-ketoglutarate. D is unlikely, since changes in isocitrate dehydrogenase activity would not be expected to affect the levels of pyruvate.

A4–25.   E. A is untrue as electrons from $FADH_2$ can be used as well. B is untrue as the two mechanisms of coupling oxidation to phosphorylation are quite different. Oxidative phosphorylation involves the oxidation of NADH to NAD^+ by proteins of the electron transport chain. Electron transport then causes the formation of a proton gradient across a membrane, which drives ATP synthesis. In contrast, glyceraldehyde 3-phosphate dehydrogenase action involves the reduction of NAD^+ to NADH and uses the $\Delta G°$ of glyceraldehyde oxidation to form a high-energy bond that can be attacked directly by a phosphate group. C is untrue, as electron transport occurs in the plasma membrane of procaryotes. D is untrue, as molecular oxygen acts as an acceptor, not a donor, for electrons.

A4–26.   E. If the inner mitochondrial membrane became permeable to protons, the electron transport chain would continue to oxidize NADH to NAD^+, transport electrons and pump protons, so the consumption of oxygen would not fall (D). However, the energy stored by the protons would be immediately dissipated as heat when they flowed back across the membrane and they could not drive the synthesis of ATP. But, NADH would not build up (B), and the citric acid cycle and glycolysis would continue (and thus CO_2 would still be produced). Since glycolysis and the citric acid cycle produce 2 molecules of ATP and one molecule of GTP (which can be converted to ATP), respectively, ATP production would not completely cease (A), but it would be very much less than normal.

A4–27.   D, both are storage polymers of glucose. A is false, since both plants and animals can also store food as fats and oils. B is false, as although glycogen synthesis requires ATP, no ATP is generated by its hydrolysis to monomers (so the number of branch points is irrelevant to energy storage). C is false, as animal cells cannot do this. E is false, as the use of protein for energy occurs only under starvation conditions.

A4–28.   B. The breakdown of glycogen to glucose does not require oxygen and the glucose can then enter glycolysis. A is incorrect, as a gram of glycogen (wet or dry) produces less energy than a gram of fat. C is incorrect, as the water bound by glycogen is not useful in keeping the body hydrated and merely contributes to making the glycogen weigh a lot. D is incorrect, as the actual mass of glycogen required to store the same amount of energy is sixfold greater than the amount of fat. E is incorrect, as fats can be carried in the bloodstream. If the energy stored in glycogen is required by other cells, glycogen is broken down to glucose, and the glucose is then released into the bloodstream.

A4–29.   C. In oxidative metabolism, fatty acids can only be converted to acetyl CoA, which is completely oxidized to carbon dioxide through the citric acid cycle. Bacteria can therefore use them as a source of carbon atoms to replenish the citric acid cycle whereas animals cannot. A, B, and E are incorrect, since glucose and fructose can be converted to pyruvate, and hence to citric acid cycle intermediates, in both animal and bacterial cells, while D is incorrect, since carbon dioxide cannot be used as a main carbon source by either nonphotosynthetic bacteria or animal cells.

A4–30.   The CO_2 was taken in by a plant leaf cell where it became part of a sugar molecule and then a (1) starch molecule in a chloroplast. At this point the leaf was eaten by you. The starch molecule was digested to glucose which was taken up by a liver cell and stored as (2) glycogen. When required this was broken down into glucose 1-phosphate, which entered glycolysis and was eventually converted into (3) fructose 1,6-bisphosphate. Glycolysis produced (4) pyruvate, which was converted into acetyl CoA, which entered the citric acid cycle, where the acetyl group derived from pyruvate combined with oxaloacetate to form (5) citrate. This is metabolized through several intermediates in the citric acid cycle which can provide the carbon skeletons for (6) amino acid biosynthesis. The amino acid was then incorporated into the enzyme when the enzyme was synthesized.

The carbon atom entered your body between steps 1 and 2.

5 Protein Structure and Function

Questions

THE SHAPE AND STRUCTURE OF PROTEINS (Pages 134–154)
The Shape of a Protein Is Specified by Its Amino Acid Sequence (Pages 134–139)

5–1   Easy, multiple choice

Which of the following statements are true?

   A. Peptide bonds are the only covalent bonds that can link together two amino acids in proteins.
   B. The polypeptide backbone of some proteins is branched.
   C. Nonpolar amino acids tend to be found in the interior of proteins.
   D. The sequence of the atoms in the polypeptide backbone varies between different proteins.
   E. A protein chain ends in a free amino group at the C-terminus.

Proteins Fold into a Conformation of Lowest Energy (Pages 139–140)

5–2   Easy, matching/fill in blanks

For each of the following sentences, fill in the blanks with the correct word selected from the list below. Use each word only once.

   A. A newly synthesized protein generally folds up into a _____ conformation.
   B. All the information required to determine a protein's conformation is contained in its amino acid _____.
   C. On heating, a protein molecule will become _____ due to breakage of _____ bonds.
   D. On removal of urea, an unfolded protein can become _____.
   E. The final folded conformation adopted by a protein is that of _____ energy.

sequence; denatured; lowest; unstable; renatured; composition; stable; reversible; irreversible; covalent; noncovalent; highest.

5–3   Difficult, multiple choice + short answer

You wish to produce a human enzyme, protein A, by introducing its gene into bacteria. The genetically engineered bacteria make large amounts of protein A, but it is in the form of an insoluble aggregate with no enzymatic activity. Which of the following procedures would be worth trying in order to obtain soluble, enzymatically active protein? Explain your reasoning.

   A. Make the bacteria synthesize protein A at a lower rate and in smaller amounts.
   B. Dissolve the protein aggregates in urea, then dilute the solution and slowly remove the urea.

C. Treat the insoluble aggregate with a protease.

D. Make the bacteria overproduce chaperone proteins in addition to protein A.

Proteins Come in a Wide Variety of Complicated Shapes (Pages 140–141)

5–4   Easy, multiple choice

Which of the following statements about proteins are true?

A. The three-dimensional structure of a protein can usually be predicted from knowledge of its amino acid sequence.

B. Two proteins having similar amino acid sequences will often have similar shapes.

C. Proteins containing fewer than 100 amino acids cannot fold into stable structures.

D. Most proteins contain more than 2000 amino acids.

E. The detailed three-dimensional structure of a protein can usually be determined by electron microscopy.

The α Helix and the β Sheet Are Common Folding Patterns (Pages 141–145)

5–5   Easy, multiple choice

The α helix and β sheet are found in many different proteins because they are formed by:

A. hydrogen bonding between the amino acid side chains most commonly found in proteins.

B. noncovalent interactions between amino acid side chains and the polypeptide backbone.

C. ionic interactions between charged amino acid side chains.

D. hydrogen bonding between atoms of the polypeptide backbone.

5–6   Easy, short answer

(A) In the schematic diagram of a protein given in Figure Q5–6, label the three protein strands that are linked together in a β sheet with a "b".

(B) Is this β sheet parallel or antiparallel?

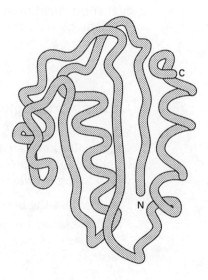

Q5–6

5–7   Intermediate, multiple choice (Requires student also to have studied Chapter 2)

Which of the following structures could the following polypeptide form? The nonpolar amino acids are italicized.

Gly-Leu-Asp-Glu-*Ile-Ala*-Lys-Ser-*Val*-Arg-His-*Phe*-Cys-His-*Ala-Ile*

    A. A hydrophobic α helix.

    B. An amphipathic α helix.

    C. A hydrophilic α helix.

    D. A hydrophobic β sheet.

    E. An amphipathic β sheet.

Proteins Have Several Levels of Organization (Pages 145–147)

5–8   Easy, matching/fill in blanks (Requires student to have studied the whole chapter)

For each of the following sentences, fill in the blanks with the correct word selected from the list below. Use each word only once.

    A. α helices and β sheets are examples of protein _____ structure.

    B. A protein such as hemoglobin, which is composed of more than one protein _____, has _____ structure.

    C. A protein's amino acid sequence is known as its _____ structure.

    D. A protein _____ is the modular unit from which many larger single-chain proteins are constructed.

    E. The three-dimensional conformation of a protein is its _____ structure.

primary; secondary; quaternary; domain; tertiary; subunit.

5–9   Intermediate, short answer (Requires information from sections on pages 167–174)

You are digesting a protein 625 amino acids long with the enzymes Factor Xa and thrombin, which are proteases that bind to and cut proteins at particular short sequences of amino acids. You know the amino acid sequence of the protein and so can draw a map of where factor Xa and thrombin should cut it (Figure Q5–9). You find, however, that treatment with each of these proteases for an hour results in only partial digestion of the protein, as summarized under the figure. List the segments (A–E) of the protein that are most likely to be folded into compact, stable domains.

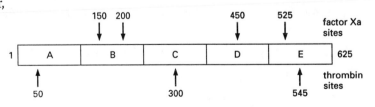

Enzyme	Sizes of fragments produced
factor Xa	100, 525
thrombin	50, 80, 245, 250

Q5–9

Few of the Many Possible Polypeptide Chains Will Be Useful (Page 147)

5–10   Easy, short answer

Calculate how many different amino acid sequences there are for a polypeptide chain 10 amino acids long.

5–11   Easy, multiple choice

All known proteins in cells adopt a single stable conformation because:

A. any chain of amino acids can fold up into only one stable conformation.

B. protein chains that can adopt several different conformations have been weeded out by natural selection.

C. chaperone proteins prevent the protein chain from adopting a preferred unstable conformation.

D. they are complexed with other molecules that keep them in that one particular conformation.

Proteins Can Be Classified into Families (Pages 147–148)

5–12   Intermediate, multiple choice + short answer (Requires information from sections on page 167 and pages 174–176)

A friend tells you that she has just discovered that the protein responsible for causing dogs to chase cars is a member of the MAP protein kinase family. In response to your blank stare, she adds that the yeast protein Ste7p, which is involved in response to a yeast hormone is also a MAP kinase family member. Although you still have no idea of what either a MAP kinase or Ste7p is, which of the following can you safely predict to be true? Explain your reasoning.

A. The dog protein and Ste7p have mostly similar amino acid sequences.

B. The dog protein and Ste7p catalyze the transfer of a phosphate group to another molecule.

C. The dog protein phosphorylates the same type of molecule that Ste7p phosphorylates.

D. The dog protein and Ste7p have identical three-dimensional structures.

E. The dog protein is involved in response to hormones.

Larger Protein Molecules Often Contain More Than One Polypeptide Chain (Pages 148–149)

5–13   Easy, multiple choice (Requires information from section on pages 171–172)

A hemoglobin molecule:

A. is composed of four protein domains.

B. is composed of four identical protein subunits.

C. is composed of four different types of protein subunits.

D. is composed of two different types of protein subunits.

E. contains one atom of iron.

Proteins Can Assemble into Filaments, Sheets, or Spheres (Pages 149–152)

5–14   Intermediate, multiple choice (Requires information from Panel 5–5)

When purified samples of protein Y and of a mutant version of protein Y are both washed through the same gel-filtration column, mutant protein Y runs through the column much slower than the normal protein. Which of the following changes in the mutant protein are most likely to explain this result?

    A. The loss of a binding site on the mutant protein surface through which protein Y normally forms dimers.

    B. A change that results in the mutant protein acquiring an overall positive instead of a negative charge.

    C. A change that results in mutant protein Y being larger than the normal protein.

    D. A change that results in mutant protein Y having a slightly different shape from the normal protein.

A Helix Is a Common Structural Motif in Biological Structures (Page 152)

5–15   Easy, multiple choice

A helical structure:

    A. will contain two, three, four, or some other exact number of subunits per each turn of the helix.

    B. that is right-handed if viewed from one end will appear to be left-handed if viewed from its other end.

    C. can form only by joining together a string of identical protein molecules.

    D. can form either within a single large molecule or from an assembly of separate molecules.

Some Types of Proteins Have Elongated Fibrous Shapes (Pages 152–154)

5–16   Easy, matching/fill in blanks

For each of the following indicate whether the individual folded polypeptide chain forms a globular (G) or fibrous (F) protein molecule.

    1. Keratin.

    2. Lysozyme.

    3. Elastin.

    4. Collagen.

    5. Hemoglobin.

    6. Actin.

Extracellular Proteins Are Often Stabilized by Covalent Cross-Linkages (Page 154)

5–17   Easy, multiple choice (Requires information from Panel 5–5)

S–S bonds:

   A. are formed by the cross-linking of methionine residues.
   B. are formed mainly in proteins that are retained within the cytosol.
   C. stabilize but do not change a protein's final conformation.
   D. can be broken by oxidation through agents such as mercaptoethanol.
   E. rarely form in extracellular proteins.

HOW PROTEINS WORK (Pages 154–179)
Proteins Bind to Other Molecules (Pages 155–156)

5–18   Easy, multiple choice

Which of the following statements are false?

   A. All protein molecules function by binding specifically to other molecules.
   B. Many proteins can bind to more than one ligand.
   C. Binding between protein and ligand generally involves noncovalent bonds.
   D. Proteins are designed to bind their ligands as tightly as possible.
   E. Changes in the amino acid sequence of a protein can decrease binding to a ligand, even if the amino acid affected by the change does not lie in the binding site for the ligand.

The Binding Sites of Antibodies Are Especially Versatile (Pages 156–157)

5–19   Intermediate, multiple choice (Requires information from Panels 5–3, 5–4, and 5–5)

You might use antibodies to purify a particular protein from a mixture using the technique of:

   A. ion-exchange chromatography.
   B. density gradient centrifugation.
   C. affinity chromatography.
   D. gel-filtration chromatography.
   E. velocity sedimentation.

Binding Strength Is Measured by the Equilibrium Constant (Pages 157–167)

5–20   Intermediate, multiple choice

The equilibrium constant for the binding of a protein to its ligand can depend on all of the following EXCEPT:

   A. the number of noncovalent bonds formed between the protein and the ligand.
   B. the concentration of the ligand.

    C. the exact fit of the binding site to the ligand.

    D. the temperature.

    E. the pH.

Enzymes Are Powerful and Highly Specific Catalysts (Page 167)

5–21   Easy, short answer (Requires student to have studied the whole chapter)

For each of the following proteins, indicate whether they have enzymatic activity (Y) or not (N).

    1. Antibody.

    2. Hemoglobin.

    3. Serine protease.

    4. Hexokinase.

    5. Motor protein.

    6. Lysozyme.

    7. GTP-binding protein.

Lysozyme Illustrates How an Enzyme Works (Pages 167–169)

5–22   Easy, matching/fill in blanks (Requires information from sections on pages 155–156 and 167)

For each of the following sentences, fill in the blanks with the correct word or phrase selected from the list below. Use each word or phrase only once.

    A. Any substance that will bind to a protein is known as its _____.

    B. Enzymes bind their _____ at the _____.

    C. Enzymes catalyze a chemical reaction by providing conditions favorable for the formation of a _____ intermediate called the _____.

    D. The enzyme hexokinase is so specific that it reacts with the D-isomer of glucose but not with the _____.

    E. Once the reaction is completed the enzyme releases the _____ of the reaction.

transition state; high-energy; low-energy; substrates; products; inhibitor; L-isomer; active site; activation energy; D-isomer; ligand.

5–23   Intermediate, multiple choice

Lysozyme catalyzes the hydrolysis of polysaccharides made of an alternating series of two related, but different amino sugars. From what you know about the mechanism of lysozyme action, which of the following statements is true?

    A. Lysozyme is most active at extremely high or extremely low pH.

    B. Hexameric polysaccharides may be cleaved into either a dimer and a tetramer or into two trimers.

C. Lysozyme is unable to cleave a very short substrate of two subunits.

D. Lysozyme is unable to cleave long substrates (more than seven subunits).

E. Lysozyme will efficiently cleave polymers of other sugars, as long as the subunits are smaller than those of the normal substrate.

V_{max} and K_M Measure Enzyme Performance (Pages 169–171)

5–24   Intermediate, short answer

You are studying two enzymes, A and B, both of which act upon substrate X, giving the products shown in Figure Q5–24A.

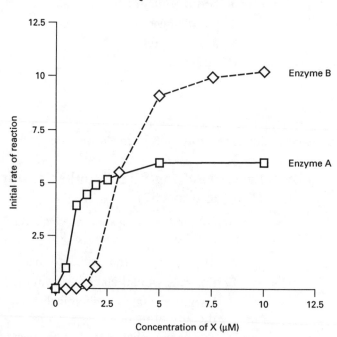

Q5–24A

You first measure the rate of the reaction X → Y in the presence of a constant amount of enzyme A and varying amounts of X. You then measure the rate of the reaction X → Z in the presence of a constant amount of enzyme B and varying amounts of X. The number of enzyme A and enzyme B molecules per milliliter (the enzyme concentrations) are exactly the same in the two experiments. The results of your studies are plotted on the graph in Figure Q5–24B.

Q5–24B

Which of the following statements are true? Explain your reasoning.

1. The turnover number of enzyme A is greater than the turnover number of enzyme B.

2. V_{max}A < V_{max}B.

3. K_MA > K_MB.

4. If both enzymes were present, at 1 μM X most of the substrate would be converted to Y rather than Z.

5–25   Intermediate, short answer (Requires information from section on pages 167–169)

Two compounds, A and B, both inhibit the same enzyme when added at a particular low concentration. At very high concentrations of the enzyme's substrate (X), the enzyme is able to overcome this inhibition. A is shaped like the enzyme's substrate, and B is shaped like the transition state molecule.

(A) What is the most likely way in which A and B inhibit the enzyme?

(B) What can you say about the relative binding affinities of A, B, and the substrate for the enzyme?

Tightly Bound Small Molecules Add Extra Functions to Proteins (Pages 171–172)

5–26   Intermediate, multiple choice

How does the prosthetic group biotin allow pyruvate carboxylase to add a carboxyl group to pyruvate?

    A. It coordinates a metal ion at the active site.

    B. It helps localize the enzyme by embedding it in the mitochondrial membrane.

    C. It absorbs light energy that is needed to drive the reaction.

    D. It forms a transient covalent bond with the carboxyl group.

    E. It lowers the K_M for the substrate.

The Catalytic Activities of Enzymes Are Regulated (Pages 172–173)

5–27   Easy, multiple choice

The biosynthetic pathway for the two amino acids E and H is shown schematically in Figure Q5–27.

You are able to show that E inhibits enzyme V, and H inhibits enzyme X. Enzyme T is most likely to be subject to feedback inhibition by:

    A. A alone.

    B. B alone.

    C. C alone.

    D. E alone.

    E. H alone.

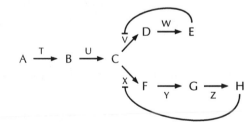

Q5–27

Allosteric Enzymes Have Two Binding Sites That Interact (Pages 173–174)

5–28   Intermediate, multiple choice

In general, a ligand that binds to only one conformation of an allosteric protein will stabilize the bound conformation. Oxygen binds to the "oxy" conformation of hemoglobin, while carbon dioxide and the small organic molecule BPG bind to the "deoxy" conformation, each at different sites. Which of the following statements are false?

    A. A high concentration of carbon dioxide will stimulate binding of BPG to hemo-
       globin.

    B. A high concentration of oxygen will stimulate dissociation of BPG and carbon
       dioxide from hemoglobin.

    C. A high concentration of carbon dioxide will cause oxygen to dissociate.

    D. In the presence of BPG, a lower concentration of carbon dioxide is required to
       cause dissociation of oxygen.

    E. A variant of hemoglobin that has a lower affinity for BPG will bind to oxygen
       less tightly than normal hemoglobin in the presence of the same level of BPG.

A Conformational Change Can Be Driven by Protein Phosphorylation (Pages 174–176)

5–29   Intermediate, multiple choice

The movement of a ciliate protozoan is controlled by a protein called RacerX. When this binds to another protein found at the base of the cilia, it stimulates the cila to beat faster and the protozoan swims faster. This ciliar protein, Speed, can be phosphorylated and only binds to RacerX in its phosphorylated form. You have identified the threonine residue at which Speed is phosphorylated and changed it to an alanine residue. How would you expect the mutant protozoan to behave?

    A. Unable to swim fast all of the time.

    B. Always swimming fast.

    C. Unable to swim fast any of the time.

    D. Switching rapidly back and forth between fast and slow swimming.

    E. Unable to move at all.

5–30   Intermediate, multiple choice

Which of the following mutant protozoa would swim fast all of the time?

    A. One lacking the protein kinase that phosphorylates the Speed protein.

    B. One lacking the Speed protein

    C. One lacking the RacerX protein.

    D. One in which the protein phosphatase that dephosphorylates the Speed pro-
       tein is produced in much greater amounts than normal.

    E. One lacking the protein phosphatase.

5–31   Difficult, short answer (Requires information from Chapter 4)

The activity of some of the enzymes of glycolysis and the citric acid cycle is regulated by phosphorylation. For these enzymes, would you expect the inactive form to be the phosphorylated form or the dephosphorylated form? Explain your answer.

GTP-binding Proteins Can Undergo Dramatic Conformational Changes (Page 176)

5–32   Intermediate, multiple choice

GTP-binding proteins:

    A. form a transient covalent bond with guanine nucleotides.

    B. are generally activated by factors that increase their rate of GTP hydrolysis.

    C. immediately release the GDP produced by GTP hydrolysis.

    D. "reset" themselves by phosphorylating bound GDP.

    E. do not readily exchange bound GDP for GTP unless stimulated to do so.

Motor Proteins Produce Large Movements in Cells (Pages 176–178)

5–33   Easy, multiple choice

A molecule of the motor protein Winnebago, supplied with ATP, is moving along a microtubule in the direction shown in Figure Q5–33.

Q5–33

What will happen if you suddenly remove all ATP from the system by adding an enzyme that hydrolyzes ATP?

    A. No change: Winnebago will continue to move from A to B.

    B. Winnebago will wander back and forth along the microtubule.

    C. Winnebago will move backwards (towards point A instead of point B).

    D. Winnebago will stall on the microtubule.

    E. Winnebago will continue to move from point A to point B, but at a slower rate.

Proteins Often Form Large Complexes That Function as Protein Machines (Pages 178–179)

5–34   Easy, multiple choice

Assembling the individual enzymes required for a multistep process into a protein machine is likely to increase the efficiency with which the entire process is carried out in all of the following ways EXCEPT:

    A. by increasing the rate at which the individual enzymes encounter their substrates.

    B. by ordering the reactions sequentially.

    C. by increasing the V_{max} of the individual enzymes.

    D. by coordinating the regulation of the individual enzymes.

    E. by coordinating the movement of the enzymes.

Answers

A5–1.   C. A is untrue, as some proteins also contain covalent disulfide bonds (–S–S– bonds) linking two amino acids. D is untrue, as the sequence of atoms in the polypeptide backbone itself is always the same from protein to protein; it is the order of the amino acid side chains that differs. E is untrue, as a protein chain has a carboxyl group at its C-terminus.

A5–2.   A. A newly synthesized protein generally folds up into a <u>stable</u> conformation.

B. All the information required to determine a protein's conformation is contained in its amino acid <u>sequence</u>.

C. On heating, a protein molecule will become <u>denatured</u> due to breakage of noncovalent bonds.

D. On removal of urea, an unfolded protein can become <u>renatured</u>.

E. The final folded conformation adopted by a protein is that of <u>lowest</u> energy.

A5–3.   A, B, and D are all worth trying. Some proteins require molecular chaperones in order to fold properly within the environment of the cell. In the absence of chaperones, the partly folded chains have a tendency to aggregate improperly with each other and with other proteins. Since the protein you are expressing in bacteria is being made in large quantities, it is possible that there are not enough chaperone molecules in the bacterium to fold the protein. If this is the case, expressing the protein slowly and at lower levels (A), or overexpressing chaperone proteins (D) might increase the amount of properly folded protein. In theory, urea should solubilize the protein and completely unfold it. Removing the urea allows the protein to refold; presumably under less crowded conditions, the protein should be able to refold into its proper conformation (B). Treating the aggregate with a protease, which cleaves peptide bonds, will probably solubilize the protein by trimming it into pieces that do not interact as strongly with one another; however, chopping the protein up will also destroy its activity.

A5–4.   B. Since a protein's three-dimensional structure is determined by its amino acid sequence, proteins with similar amino acid sequences will often have very similar shapes.

A5–5.   D.

A5–6.   (A) Figure A5–6. (B) Antiparallel.

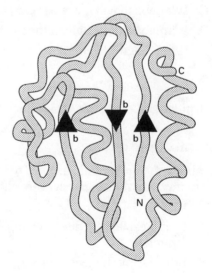

A5–6

A5–7.   B. If the sequence is twisted into an α helix, which has nearly 7 amino acids in two complete turns, the nonpolar (hydrophobic) residues will all be on one side of the helix, which will thus be amphipathic (hydrophobic on one side and hydrophilic on the other). If every other amino acid were hydrophobic, the polypeptide could form an amphipathic β sheet (E) , with alternating side chains projecting above and below the sheet. If all the amino acids were nonpolar, the polypeptide could form either a hydrophobic α helix (A) or a hydrophobic β sheet (D). Similarly, if all the side chains were polar or charged, a polypeptide could form a hydrophilic α helix.

A5–8.   A. α helices and β sheets are examples of protein <u>secondary</u> structure.

B. A protein such as hemoglobin, which is composed of more than one protein <u>subunit</u>, has <u>quaternary</u> structure.

C. A protein's amino acid sequence is known as its <u>primary</u> structure.

D. A protein <u>domain</u> is the modular unit from which many larger single-chain proteins are constructed.

E. The three-dimensional conformation of a protein is its <u>tertiary</u> structure.

A5–9.   B and D. To cut the protein chain, Factor Xa and thrombin must bind to their preferred cutting sites. If these sites are folded into the interior of a stable protein domain, it will be much more difficult for the proteases to gain access to them than if they are part of a relatively unstructured part of the chain. Hence, sites that are folded inside of a protein domain are protected from cleavage by a protease. From the sizes of the fragments produced by digestion of the protein with Factor Xa, we can conclude that the enzyme does not cut at the sites in regions B or D, although it does cut in region E. From the sizes of the fragments produced by thrombin, we can conclude that this enzyme cuts at the sites in A, C, and E. Therefore, the segments of the protein that are most likely to be folded into compact stable domains are B and D.

A5–10.   The quantity 20^{10} = approximately 10^{13}.

A5–11.   B. The other statements are untrue.

A5–12.   A and B. Members of the same protein family have similar protein sequences, similar three-dimensional structures, and roughly similar chemical activities (for example, all of the serine proteases catalyze the cleavage of a peptide bond). So if the dog protein is a MAP protein kinase, it is similar in sequence to other MAP kinases (A) and most likely has kinase activity (the transfer of a phosphate group from ATP to another molecule) (B). However, the actual substrates and the physiological function of proteins in the same family can differ quite markedly, so it is unlikely that the dog protein phosphorylates the same type of molecule that Ste7p does or is involved in the same type of response that Ste7p mediates.

A5–13.   D.

A5–14.   A. The dimers formed by the normal protein will run through the gel-filtration column faster than the mutant protein Y. B is unlikely, as gel-filtration columns separate proteins on the basis of size, not charge. C is unlikely, since if the mutant protein were larger than normal it would be less able to enter the porous beads and would run through the column at a faster speed than the normal protein. D is unlikely, as a small change in shape, without a change in size, would be unlikely to have a major effect.

A5–15.   D.

A5–16.   1= F; 2 = G; 3 = F; 4 = F; 5 = G; 6 = G.

A5–17.   C. A is incorrect since S–S bonds are formed between cysteines. B is incorrect, as they are formed mainly in extracellular proteins. D is incorrect; they are broken by mercaptoethanol, but by reduction not oxidation. E is incorrect for the reason stated in B.

A5–18.   D. Most proteins need to release their ligands at some point. For example, hemoglobin would be useless as a carrier of oxygen if it could never release the oxygen to the tissues that need it. Enzymes could be permanently disabled by feedback inhibitors if they bound to them as tightly as some antibodies bind their ligands.

A5–19.   C. An antibody that specifically binds the required protein would be attached to the inert matrix in a chromatography column. When the mixture is run through the column, the required protein is retained by noncovalent binding to the antibody. It can later be released from the column by a change in pH or treatment with a concentrated salt solution.

A5–20.   B. The equilibrium constant measures the strength of the interaction between a protein and its ligand and is independent of the concentration of either the protein or the ligand. The strength of the protein-ligand interaction increases as the number of noncovalent bonds between the two increases. The shape of the binding site affects the ability of the protein side chains to interact with portions of the substrate molecule. Both temperature and pH can disrupt noncovalent bonds which are also responsible for keeping the protein folded and thus functional.

A5–21.   1 and 2 do not have enzymatic activity. The others do. Serine protease cuts protein chains; lysozyme cuts a certain type of polysaccharide; hexokinase adds a phosphate group to a sugar molecule; GTP-binding proteins hydrolyse GTP; motor proteins hydrolyze ATP.

A5–22.   A. Any substance that will bind to a protein is known as its <u>ligand</u>.

B. Enzymes bind their <u>substrates</u> at the <u>active site</u>.

C. Enzymes catalyze a chemical reaction by providing conditions favorable for the formation of a <u>high-energy</u> intermediate called the <u>transition state</u>.

D. The enzyme hexokinase is so specific that it reacts with the D-isomer of glucose but not with the <u>L-isomer.</u>

E. Once the reaction is completed the enzyme releases the <u>products</u> of the reaction.

A5–23.   C. Lysozyme breaks the bonds between two sugar residues by distorting one of the sugar subunits. The protein compensates for this energetically unfavorable act by forming favorable interactions with five other subunits of the polysaccharide chain. A substrate having less than six subunits will not be able to form all of the favorable interactions with the enzyme and therefore cannot be bound and distorted as effectively by the enzyme (C). The six-unit substrate must bind so that all of its subunits lie in the active site of the enzyme and form contacts with the protein; since the sugars alternate, the hexamer will always be cleaved in the same place (in fact it cleaves between the fourth and fifth sugar in the bound hexamer) (B). Long substrates, on the other hand can be cleaved perfectly well (D), although the extra subunits do not contribute anything more to the binding energy. A substrate of two subunits can fit into the active site of the enzyme, but will not be able to form all of the same contacts that the true substrate forms and therefore will not bind and be cleaved efficiently (E). Lysozyme requires a protonated glutamic acid residue and a deprotonated aspartate residue in order to function properly. At very high pH, both residues will be deprotonated and at very low pH both residues will be protonated, and thus the enzyme will not function in either of these conditions (A).

A5–24.   Options 2 and 4. From the graph we can see that the V_{max} (which is directly related to the turnover number) for A is about 6 and the V_{max} for B is about 10. So the V_{max} and the turnover number are higher for B than for A (1 is thus untrue). At 1 μM X, the reaction catalyzed by A is much faster than the reaction catalyzed by B, so at this concentration, most of the substrate will be converted to Y rather than Z. Option 3 is untrue, since for A, K_M (the concentration at which the rate of the reaction is $1/2$ V_{max}) is about 0.5 µM, and for B, K_M is about 2.5 µM.

A5–25.   (A) Because the compounds are shaped like the substrate and the transition state, they are most likely to inhibit the enzyme by binding to the active site and preventing the true substrate from binding. (B) Since very high concentrations of substrate are required to overcome the inhibition by either A or B, both A and B must bind to the enzyme more tightly than does the substrate.

A5–26.   D. Biotin functions by forming a covalent intermediate with a carboxyl group. A, B, and C are functions provided by heme, lipid modifications, and retinal, respectively.

A5–27.   C. If E alone inhibited T, then it would be possible to shut down the pathway even if H were in low abundance. Likewise, if H alone inhibited T, it would be possible to shut down the pathway even if E were in low abundance. An enzyme is generally not inhibited by its substrate. Levels of C will build up only if both E and H are abundant and have inhibited V and X. It is more likely that C alone rather than B alone will inhibit T, since B will accumulate only after C has done so.

A5–28.   E. Since carbon dioxide and BPG stabilize the form of hemoglobin that oxygen cannot bind to, either BPG or carbon dioxide will stimulate the dissociation of oxygen; likewise, a high concentration of oxygen will stimulate the dissociation of carbon dioxide and BPG. Since BPG and carbon dioxide both bind to and stabilize the same form of hemoglobin, these two ligands should help each other bind. Since BPG binding stimulates the dissociation of oxygen, a variant of hemoglobin that does not bind BPG well should bind more tightly to oxygen.

A5–29.   C. The alanine side chain has no hydroxyl (OH) group and therefore cannot be phosphorylated. The altered Speed protein would therefore be expected to behave like a protein that is not phosphorylated at all times. Such a protein would never be bound to RacerX and the mutant organisms would never swim fast.

A5–30.   E. The lack of the protein phosphatase would mean that the Speed protein could remain phosphorylated all the time, causing the organism to swim fast. A mutant missing racerX (C) would not be able to swim fast at all and neither would one missing the protein kinase that phosphorylates Speed (A) or one lacking the Speed protein (B). One that overproduced the protein phosphatase (D) would keep the Speed protein permanently dephosphorylated and thus would also be unable to swim fast.

A5–31.   In general the inactive form is the phosphorylated form. The main purpose of glycolysis and the citric acid cycle is to generate ATP; thus the enzymes are inactive when ATP is high and active when ATP is low. It makes sense that cells would not want to have to phosphorylate their enzymes to turn them on when ATP levels are already low, since phosphorylation requires ATP.

A5–32.   E. GTP-binding proteins generally hydrolyze GTP and then retain the bound GDP until stimulated to exchange GDP for GTP by some other stimulus. The conformational change driven by hydrolysis, therefore, is due to the loss of a single phosphate group, not the whole guanine nucleotide (C). G proteins do not form covalent intermediates with either GTP or GDP (A). Since the GTP-bound form is usually the active form, a factor that stimulates hydrolysis will inhibit the protein (B). G proteins are reset by nucleotide exchange, not by rephosphorylation of bound GDP (D).

A5–33.   D. Motor proteins that are capable of unidirectional movement require ATP (or GTP) hydrolysis to drive them in one direction. This is because the conformational change is coupled to ATP
hydrolysis, such that movement in the reverse direction requires ATP synthesis. Hence, if ATP
is depleted, the protein will completely stop moving, being unable to move forward for lack of
ATP and unable to move backward because ATP synthesis is thermodynamically unfavorable.

A5–34.   C. If the product of one enzyme is the substrate for another, assembling the enzymes into a
machine will bring the enzymes closer to their substrates because the product will have to diffuse only a short distance to the next enzyme, and if the enzymes are properly arranged spa-
tially, one can easily imagine how the machine could also facilitate the ordering of reactions.
Gathering the enzymes into a complex also makes it easier to regulate and move all of the
enzymes together. A machine is only as good as each of its parts, however, and if an enzyme
has a low turnover number, complexing it with other proteins is unlikely to increase the
enzyme's ability to convert substrate into product.

6 DNA

Questions

THE STRUCTURE AND FUNCTION OF DNA (Pages 184–189)
Genes Are Made of DNA (Page 185)

6–1 Easy, matching/fill in blanks

Using words selected from the list below, fill in the blanks in the following brief description of the experiment using *Streptococcus pneumoniae* that determined which biological molecule carried the genetic information.

"Cell-free extracts from _____ cells of *S. pneumoniae* were fractionated into separate fractions each containing either _____ DNA, RNA, protein, or some other cell component. Each fraction was then mixed with _____ cells of *S. pneumoniae*, and its ability to change them into _____ cells resembling the _____ cells in their properties was tested by injecting the mixture into mice. Only the fraction containing _____ was able to ____the _____ cells to _____ cells, which could kill the mice."

R strain; DNA; pathogenic; transform; nonpathogenic; purified; RNA; protein; S strain; lipid; carbohydrate; unpurified.

A DNA Molecule Consists of Two Complementary Chains of Nucleotides (Pages 185–188)

6–2 Easy, multiple choice

In a DNA double helix:

    A. the two DNA strands are identical.

    B. purines pair with purines.

    C. thymine pairs with cytosine.

    D. the two DNA strands run antiparallel.

    E. the nucleotides are ribonucleotides.

6–3 Easy, art labeling

On the diagram of a small portion of a DNA molecule in Figure Q6–3, match the labels below to the numbered label lines.

    A. Base.

    B. Sugar.

    C. Phosphate.

    D. Hydrogen bond.

    E. 5′ end.

    F. 3′ end.

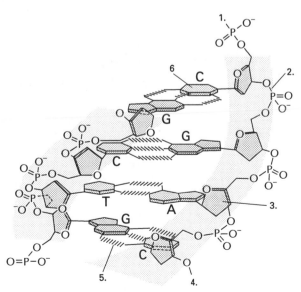

Q6–3

6–4   Easy, short answer

The structures of the four bases in DNA are given in Figure Q6–4.

(A) Which are the purines and which are the pyrimidines?

(B) Which base-pairs with which in DNA?

Q6–4

6–5   Intermediate, short answer

Using the structures in Figure Q6–4 as a guide, sketch the hydrogen bonds that form when the appropriate bases form base pairs in DNA.

6–6   Easy, multiple choice

A nucleotide sequence is said to be palindromic if it is complementary to itself. Which of the following sequences are palindromic?

    A. 5′-AAGCCGAA-3′

    B. 5′-AAGCCGTT-3′

    C. 5′-AAGCGCAA-3′

    D. 5′-AAGCGCTT-3′

    E. 5′-AATTGGCC-3′

6–7   Intermediate, multiple choice

The polarity of a strand of DNA is determined by:

    A. the sequence of bases on the strand.

    B. the presence of a charged phosphate at one end of the molecule but not the other.

    C. whether it is single stranded or is half of a double helix.

    D. the way in which the nucleotide subunits are linked to each other.

    E. the way in which the bases are linked to the sugar molecules.

6–8 **Intermediate, multiple choice (Requires information from sections on pages 201–205)**

The DNA from two different species can often be distinguished by a difference in:

    A. the ratio of A+T to G+C.

    B. the ratio of A+G to C+T.

    C. the ratio of sugar to phosphate.

    D. the presence of bases other than A, G, C, and T.

6–9 **Easy, multiple choice (Requires information from sections on pages 43–52 and 190–191)**

The first effect observed on heating a solution of double-stranded DNA to an increasingly high temperature will be:

    A. breakdown of the strands into individual nucleotides.

    B. removal of bases from the sugar-phosphate backbone.

    C. separation of the two strands.

    D. a change in the sequence of nucleotides.

6–10 **Intermediate, short answer**

Which of the two double-stranded DNAs shown below will be most resistant to the effects of heating, and why?

    A. 5′-AGCAGTTCATTATTCTCTCGTCGA-3′

       3′-TCGTCAAGTAATAAGAGAGCAGCT-5′

    B. 5′-TCCTCGAGCCTCCTGCGCCGCCGG-3′

       3′-AGGAGCTCGGAGGACGCGGCGGCG-5′

The Structure of DNA Provides a Mechanism for Heredity (Pages 188–189)

6–11 **Easy, short answer**

Given the sequence of one strand of a DNA helix:

5′-GCATTCGTGGGTAG-3′

give the sequence of the complementary strand and label the 5′ and 3′ ends.

6–12   Intermediate/difficult, short answer

(A) In principle, what is the minimum number of consecutive nucleotides that could be used to correspond to a single amino acid to give a workable genetic code? Assume that each amino acid is encoded by the same number of nucleotides. Explain your reasoning.

(B) For the nucleotide sequence CGATTG, how often, on average, would that sequence occur in a DNA strand 4000 bases long? Explain your reasoning.

DNA REPLICATION (Pages 189–198)

6–13   Easy, multiple choice

DNA replication is considered semiconservative because:

  A. after many rounds of DNA replication, the original DNA double helix is still intact.
  B. each daughter DNA molecule consists of two new strands copied from the parent DNA molecule.
  C. each daughter DNA molecule consists of one strand from the parent DNA molecule and one new strand.
  D. new DNA strands must be copied from a DNA template.

DNA Synthesis Begins at Replication Origins (Pages 190–191)

6–14   Intermediate, multiple choice (Requires information from section on pages 191–193 and requires information from Chapter 1)

If the genome of the bacterium *E. coli* requires about 20 minutes to replicate itself, how can the genome of the fruit fly *Drosophila* be replicated in only three minutes?

  A. The *Drosophila* genome is smaller than the *E.coli* genome.
  B. Eucaryotic DNA polymerase synthesizes DNA at a much faster rate than procaryotic DNA polymerase.
  C. The nuclear membrane keeps the *Drosophila* DNA concentrated in one place in the cell, which increases the rate of polymerization.
  D. *Drosophila* DNA contains more origins of replication than *E. coli* DNA.
  E. Eukaryotes have more than one kind of DNA polymerase.

New DNA Synthesis Occurs at Replication Forks (Pages 191–193)

6–15   Easy, short answer

In a DNA strand that is being synthesized, which end is growing—the 3′ end, the 5′ end, or both ends? Explain your answer.

The Replication Fork Is Asymmetrical (Pages 193–194)

6–16   Easy, multiple choice

If DNA strands were paired in a parallel rather than antiparallel fashion, how would the replication of the DNA differ from that of normal double-stranded DNA?

    A. Replication would not be semiconservative.

    B. Replication origins would not be required.

    C. The replication fork would not be asymmetrical.

    D. The polymerase used would not be self-correcting.

6–17   Intermediate, art labeling

On Figure Q6–17 of a replication bubble:

    1. Indicate where the origin of replication was located (use O).

    2. Label the leading-strand template and the lagging-strand template of the right-hand fork [R] as X and Y, respectively.

    3. Indicate by arrows the direction in which the newly made DNA strands (indicated by dark lines) were synthesized.

    4. Number the Okazaki fragments on each strand 1, 2, and 3 in the order in which they were synthesized.

    5. Indicate where the most recent DNA synthesis has occurred (use S).

    6. Indicate the direction of movement of the replication forks with arrows.

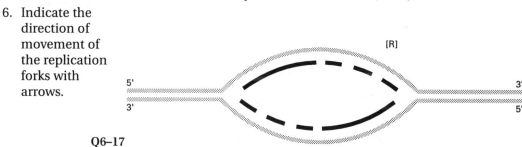

Q6–17

6–18   Easy, multiple choice

The lagging strand is synthesized discontinuously at the replication fork because:

    A. the lagging strand template is discontinuous.

    B. DNA polymerase always falls off the template DNA every ten nucleotides or so.

    C. DNA polymerase can polymerize nucleotides only in the 5′-to-3′ direction.

    D. DNA polymerase removes the last few nucleotides synthesized whenever it stops.

    E. None of the above.

DNA Polymerase Is Self-correcting (Page 194)

6–19   Intermediate, multiple choice

You have discovered a mutant form of DNA polymerase in which the proofreading function has been destroyed but the ability to join nucleotides together is unchanged. Which of the following properties do you expect the mutant polymerase to have?

A. It will polymerize in both the 5′-to-3′ direction and the 3′-to-5′ direction.

B. It will polymerize more slowly than the proofreading polymerase.

C. To replicate the same amount of DNA it will hydrolyze fewer deoxyribonucleotides than will the proofreading polymerase.

D. It will fall off the template more frequently than the proofreading polymerase.

Short Lengths of RNA Act as Primers for DNA Synthesis (Pages 194–196)

6–20   Easy, multiple choice

Replication of DNA requires a primer to initiate DNA synthesis because:

A. DNA polymerase can only add its first nucleotide to RNA.

B. DNA polymerase requires a base-paired nucleotide with a free 3′ end before it can add a new nucleotide.

C. DNA polymerase can polymerize nucleotides only in the 5′-to-3′ direction.

D. DNA polymerase can polymerize DNA only in short fragments.

6–21   Intermediate, multiple choice

RNA can be used as a primer for DNA replication in cells because:

A. primase is not found at the replication fork.

B. RNA does not base-pair to the DNA template.

C. primase synthesizes RNA in the 3′-to-5′ direction.

D. RNA spontaneously hydrolyzes after DNA polymerization has been started.

E. primase can join ribonucleotides together on a single-stranded DNA template without the need for its own primer.

6–22   Easy, multiple choice

Which of the following statements are correct?

A. Primase is less accurate than DNA polymerase at copying a DNA template.

B. The RNA primer remains as a permanent part of the new DNA molecule.

C. Replication of the leading strand does not require primase.

D. Longer primers are required to synthesize longer DNA fragments.

6–23 Intermediate, short answer

You are studying a strain of bacteria that carries a temperature-sensitive mutation in one of the genes required for DNA replication. The bacteria grow normally at the lower temperature but when the temperature is raised they die. When you analyze the remains of the bacterial cells grown at the higher temperature you find evidence of partly replicated DNA. When this DNA is denatured by heating, numerous single-stranded DNA molecules around 1000 nucleotides long are found.

(A) Which of the following proteins are most likely to be affected in these bacteria?

DNA polymerase.
Primase.
DNA ligase.
Helicase.
Initiator proteins.

(B) Explain your answer.

Proteins at a Replication Fork Cooperate to Form a Replication Machine (Pages 196–198)

6–24 Easy, multiple choice

Which of the following proteins are most abundant at the replication fork?

A. Single-strand binding protein.
B. Sliding clamp.
C. DNA polymerase.
D. Helicase.
E. Primase.

6–25 Easy/intermediate, multiple choice (Requires information from section on pages 191–193)

A molecule of bacterial DNA introduced into a yeast cell is imported into the nucleus but fails to replicate. Which of the following yeast proteins are probably unable to use the bacterial DNA as a substrate?

A. Primase.
B. Helicase.
C. DNA polymerase.
D. Sliding clamp protein.
E. Initiator proteins.

DNA REPAIR (Pages 198–205)
Changes in DNA Are the Cause of Mutations (Pages 198–200)

6–26   Intermediate, short answer

A pregnant mouse is exposed to high levels of a chemical. Many of the mice in her litter are deformed, but when they are interbred with each other, all their offspring are normal. Which TWO of the following statements could explain these results?

    A. In the deformed mice, somatic cells but not germ cells were
mutated.
    B. The original mouse's germ cells were mutated.
    C. In the deformed mice, germ cells but not somatic cells were mutated.
    D. The toxic chemical affects development but is not mutagenic.

A DNA Mismatch Repair System Removes Replication Errors That Escape from the Replication Machine (Pages 200–201)

6–27   Intermediate, multiple choice

During DNA replication in a bacterium, a C is accidentally incorporated instead of an A into one newly synthesized DNA strand. If this error is not corrected, and has no effect on the ability of the progeny to grow and reproduce, what proportion of that bacterium's progeny would you expect to contain the mutation after three more rounds of DNA replication and cell division.

    A. 100%.
    B. 50%.
    C. 25%.
    D. 10%.
    E. None.

6–28   Easy, multiple choice (Requires information from section on pages 201–202)

Mismatch repair of DNA:

    A. is carried out solely by the replicating DNA polymerase.
    B. requires an undamaged template strand.
    C. preferentially repairs the leading strand to match the lagging strand.
    D. makes replication 100,000 times more accurate.
    E. causes the condition xeroderma pigmentosum when defective.

6–29   Intermediate, multiple choice + short answer

Which of the following DNA repair processes can occur only immediately after the DNA has been replicated? Explain your answer.

    A. Repair of deamination.
    B. Repair of depurination.
    C. Mismatch repair.
    D. Repair of pyrimidine dimers.

DNA Is Continually Suffering Damage in Cells (Pages 201–202)

6–30   Easy, multiple choice

Which one of the following is the most likely explanation for a particular cancer appearing in several members of the same family?

    A. Affected individuals have inherited a cancer-causing gene that was mutated in an ancestor's somatic cells.
    B. Affected individuals have inherited a defective gene whose product interferes with DNA synthesis.
    C. Affected individuals have inherited a defective gene whose product interferes with mismatch repair.
    D. Affected individuals have inherited a defective gene whose product interferes with the synthesis of purine nucleotides.

6–31   Easy, multiple choice

If uncorrected, deamination of cytosine in DNA is most likely to lead to:

    A. substitution of an AT base pair for a CG base pair.
    B. loss of the altered CG base pair from the DNA.
    C. conversion of the DNA into RNA.
    D. generation of a thymine dimer.
    E. None of the above.

The Stability of Genes Depends on DNA Repair (Pages 202–205)

6–32   Intermediate, multiple choice

Which of the following compounds are likely to be the most mutagenic?

    A. One that depurinates DNA.
    B. One that replaces adenine with guanine.
    C. One that nicks the sugar-phosphate backbone.
    D. One that causes thymine dimers.
    E. One that partially separates the double helix.

6–33   Intermediate, short answer

You have made a collection of mutant fruit flies that are defective in various aspects of DNA repair. You test each mutant for its hypersensitivity to three DNA-damaging agents: sunlight, nitrous acid (which causes deamination of cytosine), and formic acid (which causes depurination). The results are summarized in Figure Q6–33, where a "yes" indicates that the mutant is more sensitive than a normal fly and blanks indicate normal sensitivity.

(A) Which mutant is most likely to be defective in the repair polymerase?

(B) What aspect of repair is most likely to be affected in the other mutants?

	sunlight	nitrous acid	formic acid
Dracula	yes		
Faust		yes	
Mole	yes		
Mr Self-destruct	yes	yes	yes
Marguerite			yes

Q6–33

The High Fidelity with Which DNA Is Maintained Means That Closely Related Species Have Proteins with Very Similar Sequences (Page 205)

6–34   Easy, multiple choice

You are examining the DNA sequences that code for the enzyme phosphofructokinase in skinks and Komodo dragons. You notice that the sequence that actually encodes the enzyme is very similar in the two organisms but that the noncoding sequences flanking it vary quite a bit. What is the most likely explanation for this?

A. Coding sequences are repaired more efficiently.

B. Coding sequences are replicated more accurately.

C. Coding sequences are packaged more tightly in the chromosome to protect them from DNA damage.

D. Mutations in coding sequences are more likely to be deleterious to the organism than mutations in noncoding sequences.

Answers

A6–1.   "Cell-free extracts from <u>S strain</u> cells of *S. pneumoniae* were fractionated into separate fractions each containing either <u>purified</u> DNA, RNA, protein, or some other cell component. Each fraction was then mixed with <u>R strain</u> cells of *S. pneumoniae* and its ability to change them into <u>pathogenic</u> cells resembling the <u>S strain</u> cells in their properties was tested by injecting the mixture into mice. Only the fraction containing <u>DNA</u> was able to <u>transform</u> the <u>R strain</u> cells to <u>S strain</u> (*or* <u>pathogenic</u>) cells, which could kill the mice."

A6–2.   D.

A6–3.   A, 6; B, 3; C, 2; D, 5; E, 1; F, 4.

A6–4.   (A) Cytosine and thymine are purines; adenine and guanine are pyrimidines. (B) Cytosine pairs with guanine and adenine with thymine.

A6–5.   Figure A6–5.

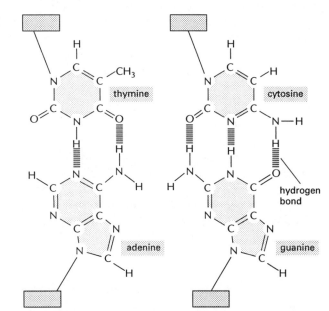

A6–5

A6–6.   D.   5'-AAGCGCTT-3'
         3'-TTCGCGAA-5'

None of the other sequences will base-pair with themselves. This double-stranded DNA molecule has the same sequence whether read forward or backward, which is why it is described as palindromic.

A6–7.   D. The polarity of a DNA strand is designated by the way in which the sugars of the sugar-phosphate backbone are linked together. Hence, a strand has polarity regardless of whether it is paired with another strand of DNA (C) and regardless of the identity of the bases (A) or the way they are attached to the backbone (E). Because each sugar subunit has a 3' end and a 5' end, a strand also retains its polarity if the phosphate on the 5' end of the strand is removed (B).

A6–8.   A. Since the sequence of nucleotides in the DNAs from different species varies considerably, but A must always pair with T and G with C, the ratio of these two base pairs will vary from species to species. B is incorrect: this ratio will always be 1 because there must always be the same amounts of A and T, and of C and G. Answer C is incorrect, since the molar ratio of sugar to phosphate is 1:1 in any DNA. D is incorrect. Although unusual bases may sometimes be formed as a result of random chemical change, this occurs in all cells, and such damage is usually rapidly repaired. A few viruses contain modified bases in their DNA; however, this is very rare.

A6–9.   C. The relatively weak hydrogen bonds holding the two strands together will break long before the much stronger covalent bonds holding the nucleotides together.

A6–10.   B. It has a higher proportion of GC base pairs (19/24) than molecule A (10/24). The higher the GC content of DNA, the more stable it is to heat denaturation, since a GC base pair forms three hydrogen bonds as opposed to an AT base pair's two.

A6–11.   5′ CTACCCACGAATGC 3′.

A6–12.   (A) Since there are 20 amino acids used in proteins, each amino acid would have to be encoded by a minimum of three nucleotides. For example, a code of two consecutive nucleotides could specify a maximum of 16 (4^2) different amino acids, excluding stop and start signals. (B) Since 4^6 (= 4096) different sequences of six nucleotides can occur in DNA, any given sequence of six nucleotides would occur on average once in a DNA strand 4000 bases long.

A6–13.   C. A and B are false. Although D is a correct statement it is not by itself the reason that DNA replication is semiconservative.

A6–14.   D. Bacteria have one origin of replication and *Drosophila* has many. A is incorrect because the *Drosophila* genome is bigger than the *E. coli* genome. B is incorrect, as eucaryotic polymerases are not faster than procaryotic polymerases. C is incorrect, since the nucleus is the same size or larger than a bacterial cell. E is incorrect, as both eucaryotes and procaryotes have more than one kind of DNA polymerase, and this is irrelevant as only two molecules of DNA polymerase can be replicating a replication fork at the same time, no matter how many types of polymerase there are.

A6–15.   The 3′ end. DNA polymerase can add nucleotides only to the 3′ end of a DNA chain.

A6–16.   C. The antiparallel nature of the strands of normal DNA requires that the replication fork be asymmetrical.

A6–17.   Figure A6–17.

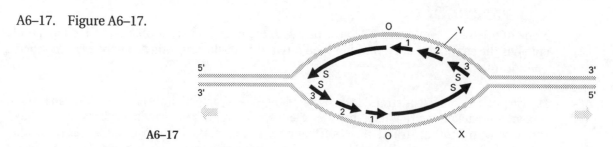

A6–17

A6–18.   C. A, B, and D are false statements.

A6–19. C. A polymerase that cannot proofread hydrolyzes fewer dNTPs than one that can because it does not check the last nucleotide added and remove it if it is incorrect. A is incorrect, as in order to polymerize in the 3′-to-5′ direction, the polymerase would have to gain an entirely new function. The rate of polymerization (B), if anything might be faster for the mutant polymerase, since it is not continually stopping to check itself.

A6–20. B. A and D are false statements. C is true but is not the reason DNA polymerase requires a primer.

A6–21. E is correct. A, B, C, and D are false statements.

A6–22. A. B is incorrect, as the RNA is removed and replaced by DNA. C is incorrect, as the leading strand requires a primer at the very beginning of the strand. D is incorrect, as the length of the primer does not correlate with the length of the DNA to be made.

A6–23. (A) DNA ligase. (B) An inactive DNA ligase would mean that Okazaki fragments formed on the lagging strand could not be joined up to form a continuous DNA and would remain as fragments. If any of the other proteins in the list had been defective, it is unlikely that any DNA replication would have occurred at all.

A6–24. A. Single-strand binding protein is required to coat all single-stranded regions of DNA that form at a replication fork; this requires many molecules of single-strand binding protein at each fork. Only one or two molecules of each of the other proteins are required at each replication fork.

A6–25. E. DNA from all organisms is chemically identical except for the sequence of nucleotides. The proteins listed in A through D can act on any DNA regardless of its sequence. In contrast, the initiator proteins recognize specific DNA sequences at the origins of replication. These sequences differ between bacteria and yeast.

A6–26. A or D. B could not account for these results since a mutation in the original mouse's germ cells would have no effect on the fetuses she was already carrying. Neither could C, as mutations in the germ cells of the fetuses while in the uterus would have had no effect on their development, but might have led to mutant mice among their offspring.

A6–27. C. DNA replication in the original bacterium will create two new DNA molecules, one of which will now carry the mismatched C. So one daughter cell of that cell division will carry a completely normal DNA molecule; the other cell will have the molecule with the mismatched C. At the next round of DNA replication and cell division, the bacterium carrying the mismatched C will produce and pass on one normal DNA molecule, from the undamaged strand, and one mutant DNA molecule, now carrying a CG pair instead of an AT pair. So at this stage, one out of the four progeny of the original bacterium is mutant. Subsequent cell divisions of these mutant bacteria will give rise only to mutant bacteria, while the other bacteria will give rise to normal bacteria.

A6–28. B. A is incorrect, as mismatch repair requires specialized repair proteins that act after the DNA is replicated. C is incorrect: it does not repair the leading strand to match the lagging strand. D is incorrect: mismatch repair makes replication approximately 100 (not 100,000) times more accurate. E is incorrect: xeroderma pigmentosum is caused by a defect in another type of DNA repair—the nucleotide excision repair system.

A6–29.   C. Mismatch repair. Mismatches occur as a result of replication errors, and thus only on the newly synthesized strand. In most cells, their repair is thought to require recognition of a newly synthesized DNA strand by nicks left in the sugar-phosphate backbone. These are sealed soon after replication, so mismatch repair must occur in the short interval between passage of the replication fork and sealing of the sugar-phosphate backbone. All the other repair processes repair damage that may occur at any time during a cell's life and to either strand of the DNA. Thus their repair processes do not depend on recognizing newly synthesized DNA strands.

A6–30.   C. A is incorrect, as mutations in somatic cells are not inherited. B and D are incorrect, as a defect in DNA synthesis or nucleotide biosynthesis itself would likely be lethal. D is incorrect, since cancer requires the accumulation of several different mutations. It would thus be extremely unlikely that several individuals in the same family spontaneously acquired exactly the same random mutations leading to the same cancer.

A6–31.   A.

A6–32.   B. Since guanine is a natural base found in DNA, the DNA repair system will be unable to recognize which strand of DNA is mutant and will repair the mismatch incorrectly half of the time. All of the other defects are easily and efficiently repaired, except E, which does not affect the sequence of the DNA.

A6–33.   (A) Mr. Self-destruct, since the repair polymerase is required for repairing most types of damage. (B) The other mutants are all most likely to be defective in the first stage of repair, the processes that excise the damaged bases. These are carried out by different proteins for the different types of damage, and so a mutant hypersensitive to sunlight, for example, will not be hypersensitive to other types of DNA damage. None of the mutants are likely to be defective in the DNA ligase step, as, like the repair polymerase, this is required for all types of repair.

A6–34.   D. Mutations—whether they have arisen by mistakes in replication or by damage to the DNA that has gone unrepaired—tend to hit the DNA fairly randomly. However, if the mutation occurs in a protein-coding sequence, it is more likely to cause a deleterious change that kills the organism, preventing the mutation from being passed on to future generations. Since skinks and Komodo dragons share a common lizard ancestor, differences in their genomes have arisen since the time they diverged from this ancestor. Mutations in noncoding sequences are more likely to have no effect and thus get passed on. A and B are incorrect because the repair and replication enzymes work on any DNA regardless of sequence. C is incorrect, as genes that are being expressed actually tend to be more loosely packaged than noncoding DNA.

7 From DNA to Protein

Questions

FROM DNA TO RNA (Pages 212–224)
Portions of DNA Sequence Are Transcribed into RNA (Pages 212–213)

7–1 Easy, multiple choice

RNA in cells differs from DNA in that:

A. it contains the base uracil, which pairs with cytosine.

B. it is single-stranded and cannot form base pairs.

C. it is single-stranded and folds up into a variety of structures.

D. the nucleotides are linked together in a different way.

E. the sugar ribose contains fewer oxygen atoms than does deoxyribose.

Transcription Produces RNA Complementary to One Strand of DNA (Pages 213–215)

7–2 Easy, multiple choice (Requires information from section on pages 216–218)

Transcription is similar to DNA replication in that:

A. it requires a DNA helicase to unwind the DNA.

B. it uses the same enzyme as that used to synthesize RNA primers during DNA replication.

C. the newly synthesized RNA remains paired to the template DNA.

D. nucleotide polymerization occurs only in the 5′-to-3′ direction.

E. an RNA transcript is synthesized discontinuously and the pieces then joined together.

Signals in DNA Tell RNA Polymerase Where to Start and Finish (Pages 216–218)

7–3 Easy, multiple choice

The promoter of a bacterial gene:

A. is located within the coding region of the gene.

B. contains short stretches of nucleotides that are identical in all promoters.

C. is located hundreds of nucleotides preceding the start site for transcription.

D. is involved in termination of transcription.

E. is required for initiation of transcription.

7–4   Easy, multiple choice

The sigma subunit of bacterial RNA polymerase:

     A. contains the catalytic activity of the polymerase.

     B. remains part of the polymerase throughout transcription.

     C. recognizes promoter sites in the DNA.

     D. recognizes transcription termination sites in the DNA.

7–5   Easy/intermediate, short answer

The DNA sequence shown in Figure Q7–5 contains the start site of transcription of a bacterial gene.

     (A) Will this start site be:

          1. to the right of the two boxed sequences?

          2. to the left of the two boxed sequences?

          3. between the two boxed sequences?

     (B) Which direction (left or right) will transcription proceed?

     (C) Which of the two strands (upper or lower) will be the template strand?

```
         5'.....CCGTGGACCTACGTAC AATATA AGCTAGCCCTAGTCGAT TGTCAA CGGTACCGATCTAAT.....3'
Q7-5     3'.....GGCACCTGGATGCATG TTATAT TCGATCGGGATCAGCTA ACAGTT GCCATGGCTAGATTA.....5'
```

7–6   Intermediate, multiple choice (Requires information from section on pages 230–232)

Which of the following might decrease the transcription of only one specific gene in a bacterial cell?

     A. A decrease in the amount of sigma factor.

     B. A decrease in the amount of RNA polymerase.

     C. A mutation that introduced a stop codon into the DNA preceding the coding sequence of the gene.

     D. A mutation that introduced extensive sequence changes into the DNA preceding the transcription start site of the gene.

     E. A mutation that moved the transcription termination signal of the gene farther away from the transcription start site.

7–7   Easy, short answer

From the list below, pick THREE reasons why the primase that is used to make the RNA primer for DNA replication would not be suitable for gene transcription?

     A. Primase initiates RNA synthesis on a single-stranded DNA template.

     B. Primase can initiate RNA synthesis without the need for a base-paired primer.

     C. Primase synthesizes only RNAs of around 5 to 20 nucleotides in length.

     D. The RNA synthesized by primase remains base-paired to the DNA template.

     E. Primase is not self-correcting.

Eucaryotic RNAs Undergo Processing in the Nucleus (Pages 218–219)

7–8 Easy, art labeling

Indicate where the following processes take place by adding numbered labeling lines to the schematic diagram of a eucaryotic cell in Figure Q7–8.

1. Transcription.
2. Translation.
3. RNA splicing.
4. Polyadenylation.
5. RNA capping.

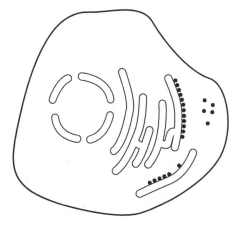

Q7–8

7–9 Easy, multiple choice

The total nucleic acids are extracted from a culture of yeast cells and are then mixed with resin beads to which the polynucleotide 5′-TTTTTTTTTTTTTTTT-TTTTTTTTTT-3′ has been covalently attached. After a short incubation, the beads are then extracted from the mixture. When you analyze the cellular nucleic acids that have stuck to the beads, which of the following will be most abundant?

A. DNA.
B. tRNA.
C. rRNA.
D. mRNA.
E. Primary transcript RNA.

Eucaryotic Genes Are Interrupted by Noncoding Sequences (Pages 219–220)

7–10 Easy, short answer

The length of a particular gene in human DNA, measured from the start site for transcription to the end of the protein-coding region, is 10,000 nucleotides, while the length of the mRNA produced from this gene is 4000 nucleotides. What is the most likely reason for this discrepancy?

Introns Are Removed by RNA Splicing (Pages 220–222)

7–11   Difficult, multiple choice + short answer

A fragment of human DNA containing the gene for a protein hormone with its regulatory regions removed is introduced into bacteria, but although it is transcribed at a high level into RNA, no protein is made. When this RNA is extracted from the bacteria, mixed with human mRNA encoding the same hormone, and then examined in the electron microscope, you see the following structure (Figure Q7–11). Which one or more of the following statements are consistent with these results? Explain your reasoning.

A. The human DNA was inserted in the bacterial DNA next to a bacterial promoter and in its normal orientation.

B. The human DNA was inserted in the bacterial DNA next to a bacterial promoter but in an orientation opposite to normal.

C. The human DNA contained an intron.

D. The human DNA acquired a deletion while in the bacterium.

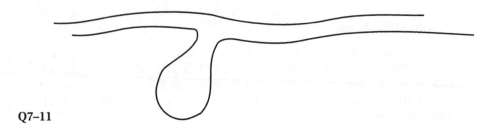

Q7–11

7–12   Intermediate, short answer

Why is the old dogma "one gene—one protein" not always true for eucaryotic genes?

The Earliest Cells May Have Had Introns in Their Genes (Pages 223–224)

7–13   Easy, multiple choice

Which of the following best describes our current model for the evolutionary history of introns?

A. They evolved to provide buffer DNA that protects eucaryotic cells from mutations.

B. They evolved to provide more origins of replication to allow eucaryotes to replicate faster.

C. They evolved to allow alternative splicing in eucaryotes.

D. They were present very early in the history of life but were lost by procaryotes.

E. They once coded for parts of proteins that could be removed without destroying protein function.

FROM RNA TO PROTEIN (Pages 224–234)
An mRNA Sequence Is Decoded in Sets of Three Nucleotides (Pages 224–225)

7–14   Easy, multiple choice

Which of the following statements about the genetic code are correct?

    A. All codons specify more than one amino acid.

    B. The genetic code is redundant.

    C. All amino acids are specified by more than one codon.

    D. The genetic code is different in procaryotes and eucaryotes.

    E. All codons specify an amino acid.

7–15   Easy, short answer (Requires information from section on pages 230–232)

The following DNA sequence includes the beginning of a sequence coding for a protein. What would be the result of a mutation that changed the C marked by an asterisk to an A?

    5′- AGGCT<u>ATG</u>AATGGACACTGCGAGCCC....
                                      *

7–16   Difficult, multiple choice

An extraterrestrial organism (ET) is discovered whose basic cell biology seems pretty much the same as terrestrial organisms except that it uses a different genetic code to translate RNA into protein. You set out to break the code by translation experiments using RNAs of known sequence and cell-free extracts of ET's cells to supply the necessary protein-synthesizing machinery. In experiments using the RNAs below, the following results were obtained when the 20 possible amino acids were added either singly or in different combinations of two or three:

RNA 1   5'-GCGCGCGCGCGCGCGCGCGCGCGCGCGCGC-3'

RNA 2   5'-GCCGCCGCCGCCGCCGCCGCCGCCGCCGCC-3'

Using RNA 1, a polypeptide was produced only if alanine *and* valine were added to the reaction mixture. Using RNA 2, a polypeptide was produced only if leucine *and* serine *and* cysteine were added to the reaction mixture. Assuming that protein synthesis can start anywhere on the template, that the ET genetic code is nonoverlapping and linear, and that each codon is the same length (like the terrestrial triplet code), how many nucleotides does an ET codon contain?

    A. 2.

    B. 3.

    C. 4.

    D. 5.

    E. 6.

tRNA Molecules Match Amino Acids to Codons in mRNA (Page 225–227)

7–17   Easy, multiple choice

Which amino acid would you expect a tRNA with the anticodon 5′-CUU-3′ to carry? The table of codons is given in Figure Q7–17.

    A. Lysine.

    B. Glutamate.

    C. Glutamine.

    D. Leucine.

    E. Phenylalanine.

Q7–17

GCA GCC GCG GCU	AGA AGG CGA CGC CGG CGU	GAC GAU	AAC AAU	UGC UGU	GAA GAG	CAA CAG	GGA GGC GGG GGU	CAC CAU	AUA AUC AUU	UUA UUG CUA CUC CUG CUU	AAA AAG	AUG	UUC UUU	CCA CCC CCG CCU	AGC AGU UCA UCC UCG UCU	ACA ACC ACG ACU	UGG	UAC UAU	GUA GUC GUG GUU	UAA UAG UGA
Ala	Arg	Asp	Asn	Cys	Glu	Gln	Gly	His	Ile	Leu	Lys	Met	Phe	Pro	Ser	Thr	Trp	Tyr	Val	stop
A	R	D	N	C	E	Q	G	H	I	L	K	M	F	P	S	T	W	Y	V	

7–18   Intermediate, multiple choice

Which of the following pairs of codons might you expect to be read by the same tRNA as a result of wobble? Codons are given in Figure Q7–17.

    A. CUU and UUU.

    B. GAU and GAA.

    C. CAC and CAU.

    D. AAU and AGU.

    E. CCU and GCU.

Specific Enzymes Couple tRNAs to the Correct Amino Acid (Page 227)

7–19   Easy, multiple choice

Which of the following steps in protein synthesis require ATP hydrolysis?

    A. Joining two amino acids together on the ribosome.

    B. Attaching an amino acid to a tRNA.

    C. Binding of tRNA to the ribosome.

    D. Binding of the ribosome to mRNA.

    E. Base-pairing of codon and anticodon.

7–20   Difficult, data interpretation + multiple choice + short answer

A strain of yeast translates mRNA into protein with a high level of inaccuracy. Individual molecules of a particular protein isolated from this yeast have the following variations in the first 11 amino acids compared with the sequence of the same protein isolated from normal yeast cells (Figure Q7–20).

What is the most likely cause of this variation in protein sequence? Explain your answer.

A. A mutation in the DNA coding for the protein.

B. A mutation in the anticodon of the isoleucine tRNA (tRNA^{Ile}).

C. A mutation in the isoleucyl-tRNA synthetase that decreases its ability to distinguish between different amino acids.

D. A mutation in the isoleucyl-tRNA synthetase that decreases its ability to distinguish between different tRNA molecules.

E. A mutation in a component of the ribosome that allows binding of incorrect tRNA molecules to the A-site.

```
normal sequence  Met Thr Ala Ile Val Ser Asn Thr Gln Ile Lys

       variants  Met Thr Ala Ala Val Ser Asn Thr Gln Ile Lys
                 Met Thr Ala Gly Val Ser Asn Thr Gln Ile Lys
                 Met Thr Ala Val Val Ser Asn Thr Gln Ile Lys
                 Met Thr Ala Ile Val Ser Asn Thr Gln Ala Lys
                 Met Thr Ala Ile Val Ser Asn Thr Gln Gly Lys
    Q7–20        Met Thr Ala Ile Val Ser Asn Thr Gln Val Lys
```

The RNA Message Is Decoded on Ribosomes (Pages 227–230)

7–21   Easy, multiple choice

Which of the following statements are correct?

A. Ribosomes are large RNA structures composed solely of rRNA.

B. Ribosomes are synthesized entirely in the cytoplasm.

C. rRNA probably contains the catalytic activity that joins amino acids together.

D. A ribosome consists of two equally sized subunits.

E. A ribosome binds one tRNA at a time.

7–22   Easy, data interpretation

Figure Q7–22A shows the stage in translation when an incoming aminoacyl-tRNA has bound to the A-site on the ribosome. Using the components shown in Figure Q7–22A as a guide, show on Figures Q7–22B and C what happens in the next two stages to complete the addition of the new amino acid to the growing polypeptide chain.

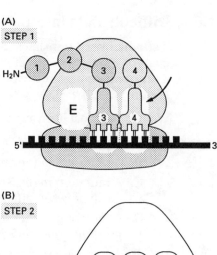

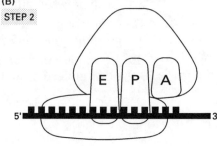

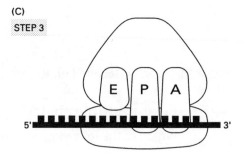

Q7–22

7–23   Intermediate, multiple choice

A poison added to an *in vitro* translation mixture containing mRNA molecules with the sequence 5′-AUGAAAAAAAAAAAAAUAA-3′ has the following effect: the only product made is a Met-Lys peptide that remains attached to the ribosome. What is the most likely way in which the poison acts to inhibit protein synthesis?

    A. It inhibits binding of the small subunit of the ribosome to mRNA.

    B. It inhibits peptidyl transferase activity.

    C. It inhibits movement of the small subunit relative to the large subunit.

    D. It inhibits release factor.

    E. It mimics release factor.

Codons in mRNA Signal Where to Start and to Stop Protein Synthesis (Pages 230–232)

7–24   Easy, multiple choice

In eucaryotes, but not procaryotes, ribosomes find the start site of translation by:

    A. binding directly to a ribosome-binding site preceding the initiation codon.

    B. scanning along the mRNA from the 5′ end.

C. recognizing an AUG codon as the start of translation.

D. binding an initiator tRNA.

7–25   Intermediate, art labeling + short answer

Figure Q7–25 shows an mRNA molecule.

(A) Match the labels given in the list below with the label lines in Figure Q7–25.

   A. ribosome-binding site.

   B. initiator codon.

   C. stop codon.

   D. untranslated 3′ region.

   E. untranslated 5′ region.

   F. protein-coding region.

(B) Is the mRNA shown procaryotic or eucaryotic? Explain your answer.

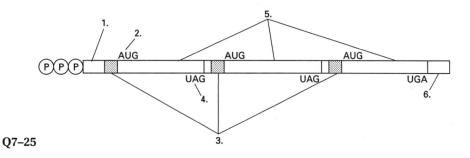

Q7–25

7–26   Difficult, multiple choice

A tRNA for the amino acid lysine is mutated such that the sequence of the anticodon is 5′-UAU-3′ (instead of 5′-UUU-3′). Which of the following aberrations in protein synthesis might this tRNA cause? The genetic code table is given in Figure Q7–17.

   A. Read through of stop codons.

   B. Substitution of lysine for isoleucine.

   C. Substitution of lysine for tyrosine.

   D. Substitution of lysine for internal methionine.

   E. Substitution of lysine for amino-terminal methionine.

Proteins Are Made on Polyribosomes (Page 232)

7–27   Easy, multiple choice

The translation of an mRNA molecule on a polyribosome:

   A. produces a single protein molecule.

   B. occurs only in procaryotes.

   C. produces many copies of the same protein.

D. occurs only in eucaryotes.

E. produces many different proteins.

7–28 Intermediate, data interpretation + multiple choice

You have discovered a protein that inhibits translation. When you add this inhibitor to a mixture capable of translating human mRNA and centrifuge the mixture to separate polyribosomes and single ribosomes, you obtain the results shown in Figure Q7–28. Which of the following interpretations are consistent with these observations?

A. The protein binds to the small ribosomal subunit and increases the rate of initiation of translation.

B. The protein binds to sequences in the 5′ region of the mRNA and inhibits the rate of initiation of translation.

C. The protein binds to the large ribosomal subunit and slows down elongation of the polypeptide chain.

D. The protein binds to sequences in the 3′ region of the mRNA and prevents termination of translation.

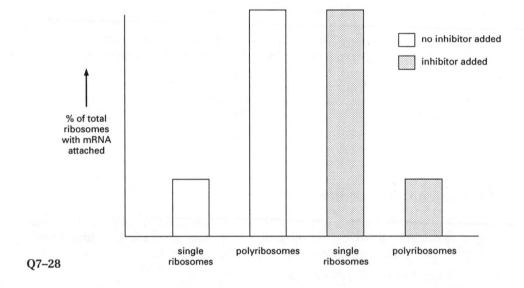

Q7–28

Carefully Controlled Protein Breakdown Helps Regulate the Amount of Each Protein in a Cell
(Pages 232–234)

7–29 Intermediate, multiple choice

The concentration of a particular protein X in a normal human cell rises gradually from a low point immediately after cell division to a high point just before cell division, and then drops sharply. The level of its mRNA in the cell remains fairly constant throughout this time. Protein X is required for cell growth and survival, but the drop in its level just before cell division is essential for division to proceed. You have isolated a line of human cells that grow in size in culture but cannot divide, and on analyzing these mutants you find that levels of X mRNA in the mutant cells are normal. Which of the following mutations in the gene for X could explain these results?

A. The introduction of a stop codon that truncates protein X at the fourth amino acid.

B. A change to the coding sequence that results in a protein X that does not fold into its normal 3D structure.

C. The deletion of a sequence that encodes sites at which ubiquitin can be attached to the protein.

D. A change at a splice site that prevents splicing of the RNA.

There Are Many Steps Between DNA and Protein (Page 234)

7–30   Easy, multiple choice

At which of the following stages in gene expression is the production of a specific protein by a cell most commonly controlled?

A. Transcription.

B. Translation.

C. mRNA degradation.

D. Protein degradation.

E. RNA splicing.

RNA AND THE ORIGINS OF LIFE (Pages 234–240)
Simple Biological Molecules Can Form Under Prebiotic Conditions (Pages 235–237)

7–31   Easy, multiple choice

Experiments simulating conditions on the early earth:

A. assume that the atmosphere at that time contained methane, ammonia, water, and abundant oxygen.

B. prove that small organic molecules such as amino acids and sugars could be synthesized without protein catalysts.

C. prove that polynucleotides originated on earth with the help of inorganic mineral catalysts.

D. have reproduced self-replicating molecules that are capable of having evolved into living cells.

7–32   Intermediate, multiple choice

According to current thinking, the minimum requirement for life to have originated on earth was the formation of:

A. a molecule that could provide a template for the production of a complementary molecule.

B. a double-stranded DNA helix.

C. a molecule that could direct protein synthesis.

D. a molecule that could catalyze its own replication.

RNA Can Both Store Information and Catalyze Chemical Reactions (Pages 237–239)

7–33   Easy, multiple choice

Which of the following reactions are known to be carried out by a ribozyme?

A. DNA synthesis.

B. Transcription.

C. RNA splicing.

D. Protein hydrolysis.

E. Polysaccharide hydrolysis.

7–34   Easy, multiple choice

You are studying a disease that is caused by a virus, but when you purify the virus particles and analyze them you find they contain no trace of DNA. Which of the following molecules are likely to contain the genetic information of the virus?

A. Protein.

B. RNA.

C. Lipids.

D. Carbohydrates.

RNA Is Thought to Predate DNA in Evolution (Pages 239–240)

7–35   Intermediate, multiple choice

A major advantage to having a DNA genome rather than an RNA genome is that:

A. RNA cannot be double-stranded.

B. a strand of DNA can encode more information than can a strand of RNA of equal length.

C. proteins cannot be made without DNA.

D. DNA contains thymine instead of uracil.

E. deoxyribose is more easily synthesized than is ribose.

Answers

A7–1.   C. A is untrue, since although RNA contains uracil, uracil pairs with adenine, not cytosine. B is untrue. RNA can form base pairs with a complementary RNA or DNA sequence. D is untrue. E is untrue. Ribose contains one more oxygen atom than deoxyribose.

A7–2.   D. RNA polymerase unwinds only a few base pairs of the double helix at a time and does not need a helicase to do so (A). The enzyme used to make primers during DNA synthesis is indeed an RNA polymerase, but it is a special enzyme, primase, and not the enzyme that is used for transcription (B). C is untrue. E is untrue: an RNA transcript is made by a single polymerase molecule that proceeds from the start site to the termination site without falling off.

A7–3.   E. A is incorrect: promoters are located before the coding region. B is incorrect: promoters are similar but not identical in DNA sequences. C is incorrect: bacterial promoters are located close to the start site of transcription. D is incorrect.

A7–4.   C. The other options are untrue statements.

A7–5.   (A) Transcription will start in the leftmost part of the sequence. The two boxed sequences indicate a bacterial promoter and tell you its orientation. The sequence

  5′-AATATA-3′

  3′-TTATAT-5′

is the nearest to the start site. (B) The orientation of the promoter determines the direction of transcription, which will be to the left in this diagram. (C) Since transcription can proceed only from 5′ to 3′, the template strand will be the upper strand.

A7–6.   D. Such changes would probably destroy the function of the promoter, making RNA polymerase unable to bind to it. Decreasing the amount of σ factor or RNA polymerase (A or B) would affect the transcription of most of the genes in the cell, not just one specific gene. Introducing a stop codon before the coding sequence (C) would have no effect on transcription of the gene, since the transcription machinery does not recognize translational stops. Moving the termination signal farther away (E) would merely make the transcript longer.

A7–7.   A, C, and D. B and E are true for both primase and RNA polymerase, and are not relevant to the point at issue.

A7–8.   Figure A7–8.

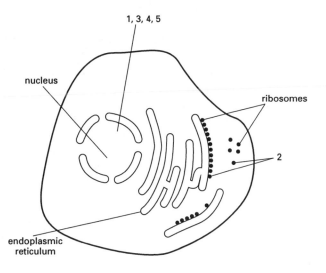

A7–8

A7–9.   D. mRNA is the only type of RNA that is polyadenylated, and this poly(A) tail would be able to base-pair with the strands of poly(T) on the beads and thus stick to them. DNA would not be found in the sample, as the poly(A) tail is not encoded in the DNA and long runs of T are rare in DNA.

A7–10.   The gene contains one or more introns.

A7–11.   B, C, and D are consistent with the results. B must be true for the RNA produced in the bacterium to be complementary to, and thus able to pair with, the mRNA from a human cell. If the human DNA had become inserted in its normal orientation next to the promoter, the corresponding portions of RNA would be identical (or at least very similar) in sequence and thus the two RNAs would not be complementary and would not pair. The loop formed in the hybrid tells us that one of the molecules contains sequences that the other is missing. This could come about either because the bacterial RNA was transcribed from human sequences that acquired a deletion or because the human gene has an intron whose sequences would be spliced out of the human mRNA and not the bacterial transcript. We cannot tell which RNA molecule is which in the electron microscope, so which of the two possibilities is the correct one cannot be determined.

A7–12.   The transcripts from some genes can be spliced in more than one way to give mRNAs containing different sequences and thus encoding different proteins. Thus, a single eucaryotic gene may encode more than one protein.

A7–13.   D. Introns are thought to have been present in early cells and to have aided in protein evolution by enhancing the rate at which the sequences coding for protein domains could be shuffled to create new genes. Procaryotes and other fast-growing organisms that benefit from having as small a genome as possible are thought to have lost introns to allow them to reproduce more quickly.

A7–14.   B. The majority of the amino acids can be specified by more than one codon. A is incorrect: each codon specifies only one amino acid. C is incorrect: tryptophan and methionine are encoded by only one codon. D is incorrect: with a few minor exceptions, the genetic code is the same in all organisms. E is incorrect: some codons specify translational stop signals.

A7–15.   The change creates a stop codon (TGA, or UGA in the mRNA) very near the beginning of the protein-coding sequence and in the correct reading frame (the beginning of the coding sequence is indicated by the ATG). Thus translation would terminate after only four amino acids had been joined together, and the complete protein would not be made.

A7–16.   D. An organism having codons with an even number of nucleotides (i.e., 2, 4, or 6) could read 5′-GCGCGCGCGC-3′ (RNA 1) in either of two ways, namely "GC GC GC GC..." or "CG CG CG CG..." So either one of the two amino acids alone could have supported protein synthesis, and you would not need them in combination (thus eliminating A, C, and E). An organism having three bases per codon could read the sequence 5′-GCCGCCGCCGCCGCC-3′ (RNA 2) in one of three ways, namely "GCC GCC GCC GCC...," "CCG CCG CCG CGG...," or "CGC CGC CGC CGC...," and so again, any one of the three amino acids could have supported synthesis of a polypeptide, and you would not need to add all three. Only a five-nucleotide code gives you two different consecutive codons for RNA 1 and three different consecutive codons for RNA 2.

A7–17.   A. Lys (lysine). As is conventional for nucleotide sequences, the anticodon is given 5′ to 3′. The complementary base-pairing occurs between antiparallel nucleic acid sequences, and the codon recognized by this anticodon will therefore be 5′-AAG-3′.

A7–18.   C. These two codons differ only in the third position and also encode the same amino acid. The codons GAU and GAA (B), although also differing only in the third position, are unlikely in normal circumstances to be read by the same tRNA, as they encode different amino acids.

A7–19.   B. The energy in the bond formed at this step between amino acid and tRNA is subsequently used to join two amino acids together at the ribosome. C, D, and E are noncovalent interactions and do not require an additional input of energy.

A7–20.   C. A mutation in the isoleucyl-tRNA synthetase that decreases its ability to distinguish between amino acids would allow an assortment of amino acids to be attached to the tRNA^{Ile}. These assorted aminoacyl-tRNAs would then base-pair with the isoleucine codon and cause a variety of substitutions at positions normally occupied by isoleucine. A is incorrect: a mutation in the gene encoding the protein would cause only a single variant protein to be made. E is incorrect: a mutation in the ribosome that allows binding of any amino-acyl-tRNA to the A site would cause substitutions all over the protein, not only at isoleucine residues. B and D are incorrect: a mutation in the anticodon loop of tRNA^{Ile} (B) or a mutation in the isoleucine-tRNA synthetase that decreases its ability to distinguish between different tRNA molecules (D) would cause substitution of isoleucine for some other amino acid (which is the opposite of what is observed).

A7–21.   C. A is incorrect: ribosomes contain proteins as well as rRNA. B is incorrect: rRNA is synthesized in the nucleus, and ribosomes are partly assembled in the nucleus. D is incorrect: a ribosome consists of one small subunit and one large subunit. E is incorrect: a ribosome must be able to bind two tRNAs at any one time.

A7–22.   Figure A7–22.

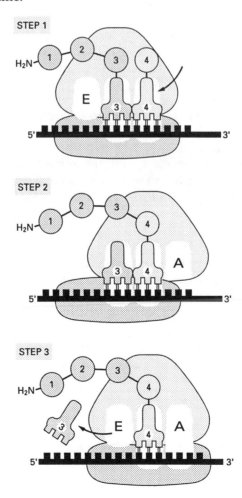

A7–22

A7–23.   C. Either A or B would prevent all peptide bond formation. D would have no affect on transla-tion until the stop codon was reached. E would be likely to result in a mixture of polypeptides of various lengths; a poison mimicking a release factor could conceivably cause only Met-Lys to be made, but this peptide would not remain bound to the ribosome.

A7–24.   B. A is true only for procaryotes. C and D are true for both procaryotes and eucaryotes.

A7–25.   (A) 1, E; 2, B; 3, A; 4, C; 5, F; 6, D. (B) The mRNA is procaryotic. It contains coding regions for more than one protein, as shown by the multiple initiation codons, each preceded by a ribo-some-binding site. It contains an unmodified 5′ end, as shown by the three phosphate groups, and an unmodified 3′ end, as shown by the absence of a poly(A) tail.

A7–26.   B and D. The mutant tRNA^{Lys} will be able to pair with the codon 5′-AUA-3′, which codes for isoleucine. Because tRNA-mRNA pairing often has "wobble" in the third codon position, the anticodon 5′-UAU-3′ should also be able to pair with the other isoleucine codons (AUC, and AUU). Because of wobble, there might also be the possibility that this mutant tRNA would be able to pair with the codon 5′-AUG-3′, which normally codes for methio-nine (D). The normal tRNA^{Ile} would not, as this would make accurate protein synthesis impossible, but since we are dealing here with a different tRNA, and the capacity for wobble base-pairing is determined by the tRNA structure, this possibility cannot be ruled out. There will be no substitution at the amino-terminal methionine codon (E) because a special initiator tRNA is used for this, and the mutant tRNA^{Lys} would not be recognized by the initiation factors.

A7–27.   C. A polyribosome consists of a single mRNA molecule that is being translated successively by numerous ribosomes, each synthesizing one copy of the protein chain.

A7–28.   B. The results in Figure Q7–28 show a marked decrease in the number of polyribosomes formed relative to normal. Polyribosomes form because the initiation of translation is fairly rapid: ribosomes can bind successively to the free 5′ end of an mRNA molecule and start trans-lation before the first ribosome has had a chance to finish translating the message. Therefore, inhibition of the rate of initiation will tend to decrease the number of ribosomes in the polyri-bosome, and in the extreme case there will be only one ribosome per mRNA. Conversely, increasing the rate of initiation or slowing the rate of elongation would result in an increased number of ribosomes per polyribosome (up to a maximum point), making A and C false. D is incorrect, as preventing termination would prevent release of the ribosomes at the end of the coding sequence and would be expected to "freeze" the assembled polyribosomes, so that the ratio of polyribosomes to ribosomes would be much as normal.

A7–29.   C. The drop in level of protein X in the normal cell is most likely due to protein degradation, since levels of mRNA remain constant. So the inability of the mutant cell to divide could be due to a mutation that inhibits protein degradation. This would be achieved by removal of sites for attachment of ubiquitin, which targets proteins for destruction. A, B, and D would probably not produce the result described, as without the production of a functional protein X the mutant cells could not grow in size.

A7–30.   A. Transcriptional regulation is the most common form of gene regulation, as one might expect, since it is the first step in gene expression.

A7–31.   B. The "prebiotic soup" experiments showed that it is possible to obtain many small organic molecules simply by mixing compounds such as methane, ammonia, and water together and exposing them to an electric current. A is incorrect, as the atmosphere of the early earth is thought not to have contained much oxygen. While they showed that polynucleotides could be formed under similarly simple conditions with the help of inorganic mineral catalysts, they do not prove that this is how polynucleotides actually originated. Self-replicating molecules have not yet been created under the conditions of these experiments.

A7–32.   D. A is incorrect in that although this may have been a step in self-replication, it would not by itself be sufficient. B and C are incorrect, as these stages in the evolution of the cell must have succeeded the formation of the first self-replicating molecules.

A7–33.   C.

A7–34.   B.

A7–35.   D. The possession of thymine instead of uracil means that when uracil is formed by spontaneous deamination of cytosine in DNA, which is a common occurrence, it can be recognized as damage and rapidly repaired. A is incorrect: two complementary strands of RNA could in principle base-pair with each other and form a double helix. B is incorrect: a strand of DNA contains the same amount of information as a strand of RNA of equal length. C is incorrect: translation per se requires only a message (mRNA) and ribosomes (protein and RNA), along with charged tRNA molecules and some other protein factors. E is incorrect, as deoxyribose has to be made by enzymatic conversion of ribose.

8 Chromosomes and Gene Regulation

Questions

THE STRUCTURE OF EUCARYOTIC CHROMOSOMES (Page 246–257)
Eucaryotic DNA Is Packaged into Chromosomes (Pages 246–247)

8–1 Easy, matching/fill in blanks

For each of the following sentences, fill in the blanks with the correct word selected from the list below. Use each word only once.

    A. In eucaryotic chromosomes, DNA is complexed with proteins to form _____.

    B. The paternal and maternal copies of human Chromosome 1 are _____.

    C. The human X and Y chromosomes are _____.

    D. Each mitotic chromosome is constricted at a site called the _____.

    E. Mitotic chromosomes are more _____ than _____ chromosomes.

kinetochore; interphase; chromatin; nonhomologous; homologous; karyotype; bands; centromere; centrosome; condensed; extended.

8–2 Intermediate, multiple choice (Requires information from sections on pages 247–253 and pages 258–259, and also from Chapter 6)

A bacterial genome differs from a typical eucaryotic genome in that the bacterial DNA:

    A. does not contain telomeres.

    B. is packaged into nucleosomes of a different size.

    C. does not contain specific origins of replication.

    D. is not complexed with proteins.

    E. does not contain gene regulatory sequences.

Chromosomes Exist in Different States Throughout the Life of a Cell (Pages 247–249)

8–3 Easy, multiple choice (Requires information from sections on pages 252–253)

Which of the following statements about eucaryotic chromosomes are true?

    A. Chromosomes are at their most condensed during interphase.

    B. The position of the centromere with respect to the chromosome ends varies in different chromosomes.

    C. The DNA becomes more condensed just before DNA replication occurs.

    D. Chromosomes are visible in the light microscope throughout the cell cycle.

    E. Most genes in mitotic chromosomes are being transcribed.

Specialized DNA Sequences Ensure That Chromosomes Replicate Efficiently (Pages 249–250)

8–4   Easy, multiple choice

What would be one of the consequences if a eucaryotic chromosome lacked telomeres?

    A. DNA replication could not be initiated efficiently.

    B. The chromosome would not attach to the mitotic spindle.

    C. The chromosome would not be replicated completely.

    D. The DNA would not become condensed.

    E. The DNA would not be transcribed.

Nucleosomes Are the Basic Units of Chromatin Structure (Pages 250–252)

8–5   Intermediate, matching/fill in blanks

For each of the following sentences, choose one of the options enclosed in square brackets to make a correct statement about nucleosomes.

    A. Nucleosomes are present in [procaryotic / eucaryotic] chromosomes, but not in [procaryotic / eucaryotic] chromosomes.

    B. Each nucleosome contains two molecules each of histones [H1 and H2A / H2A and H2B] as well as histones H3 and H4.

    C. A nucleosome core particle contains a core of histone with DNA wrapped around it approximately [twice / three times / four times].

    D. Nucleosomes are aided in their formation by the high proportion of [acidic / basic / polar] amino acids in histone proteins.

    E. Nucleosome formation compacts the DNA into approximately [one-third / one-hundredth / one-thousandth] of its original length.

Chromosomes Have Several Levels of DNA Packing (Pages 252–253)

Interphase Chromosomes Contain Both Condensed and More Extended Forms of Chromatin (Pages 253–256)

8–6   Easy, matching/fill in blanks (Requires information from section on pages 250–252)

For each of the following sentences, fill in the blanks with the correct word or phrase selected from the list below. Use each word or phrase only once.

    A. Interphase chromosomes contain both darkly staining _____ and more lightly staining _____.

    B. A string of nucleosomes coils up with the help of _____ to form the more compact structure of the _____.

    C. Genes that can be transcribed are thought to be contained in a relatively loosely packed type of euchromatin known as _____.

    D. Nucleosome core particles are separated from each other by stretches of _____ DNA.

E. After mitosis the mitotic chromatin _____ to form the interphase chromosomes in the daughter cells.

heterochromatin; euchromatin; active chromatin; folds up; mitotic chromatin; linker; 30-nm
fiber; histone H1; histone H3; histone H4; histone H2; unfolds.

8–7 Intermediate, short answer

In which of the following instances can the state of chromatin packing differ? Explain your
reasoning.

1. Between different cells of the same organism.
2. In different stages of the cell cycle.
3. In different parts of the same chromosome.
4. In different members of a pair of homologous chromosomes.

8–8 Difficult, multiple choice (Requires a basic knowledge of genetics)

If the mottled coloring of calico cats is due to X-chromosome inactivation, which of the following statements will be true?

A. Calico cats can be male or female.
B. Female calico cats will be the same colour as their mother.
C. The mottled color is due to X chromosomes repeatedly switching back and
forth between active and inactive states during development.
D. Calico cats with identical patterns will be rare.

Position Effects on Gene Expression Reveal Differences in Interphase Chromosome Packing (Page 256)

8–9 Difficult, multiple choice + short answer (Requires student to have studied the whole chapter)

A variety of peach, Desert Peach, has a uniformly fuzzy skin that is mainly white with small
random blotches of pink. The gene required for pinkness and the gene required for fuzziness
are located on the same chromosome, chromosome A. A spontaneous mutant form of this
peach, called Old Man, produces fruit with the same type of white and pink coloring but with
fuzz only where the skin is pink. Which one or more of the following statements are consistent
with the appearance of these two peaches? Explain your reasoning.

A. Old Man has a mutation that inactivates a gene encoding a gene regulatory
protein required for activation of transcription of the fuzziness gene.
B. Old Man has a mutation that inactivates a gene encoding a gene regulatory
protein required for repression of transcription of the fuzziness gene.
C. A rearrangement of chromosome A in Old Man has placed the fuzziness gene
next to the pinkness gene.
D. The gene for fuzziness is in a region of heterochromatin in Desert Peach but
not in Old Man.
E. The gene for pinkness is in a region of heterochromatin in both peaches.

8–10   Difficult, multiple choice

You have isolated a DNA-binding protein from peaches and find that when you express this protein at high levels in Desert Peach (see Question 8–9), the fruit produced are uniformly white and fuzzy. When you express this protein in Old Man, the fruit produced are uniformly white and bald. Which of the following is the most likely explanation for the function of your DNA-binding protein?

   A. It causes chromosomal condensation.
   B. It inhibits heterochromatin formation.
   C. It promotes heterochromatin formation.
   D. It is a repressor of the gene required for pinkness.
   E. It is a repressor of the gene required for fuzziness.

Interphase Chromosomes Are Organized Within the Nucleus (Page 256–257)

8–11   Easy, multiple choice

A fluorescently labeled DNA sequence that hybridizes to rRNA will stain:

   A. the nuclear envelope.
   B. the nuclear lamina.
   C. the entire nucleus.
   D. centromeres and telomeres.
   E. the nucleolus.

GENE REGULATION (Pages 257–274)
Cells Regulate the Expression of Their Genes (Pages 258–259)

8–12   Easy, multiple choice

The distinct characteristics of different cell types in a multicellular organism are produced mainly by the differential regulation of the:

   A. replication of specific genes.
   B. transcription of genes transcribed by RNA polymerase II.
   C. transcription of housekeeping genes.
   D. translation of mRNA.
   E. packing of DNA into nucleosomes in some cells and not others.

Transcription Is Controlled by Proteins Binding to Regulatory DNA Sequences (Pages 259–261)

8–13   Intermediate, multiple choice

Which of the following statements are true for homeodomain proteins?

   A. They have similar amino acid sequences.
   B. They have similar structures.
   C. They bind to the same DNA sequence.
   D. They control the same types of genes.

Repressors Turn Genes Off and Activators Turn Them On (Pages 261–263)

8–14   Easy, matching/fill in blanks

For each of the following sentences, fill in the blanks with the correct word selected from the list below. Use each word only once.

    A. The genes of a bacterial _____ are transcribed into a single mRNA.

    B. Many bacterial promoters contain a region known as an _____, to which a specific gene regulatory protein binds.

    C. Genes in which transcription is prevented are said to be _____.

    D. Bacterial genes are regulated by small molecules such as tryptophan by the interaction of such molecules with _____ DNA-binding proteins such as the tryptophan repressor.

    E. Genes that are being _____ expressed are being transcribed all the time.

operator; operon; promoter; repressed; induced; constitutively; allosteric; negatively; positively.

8–15   Intermediate, short answer

You have discovered an operon in a bacterium that is only turned on when sucrose is present and glucose is absent. You have also isolated three mutants which have changes in the upstream regulatory sequences of the operon and whose behavior is summarized below. You hypothesize that there are two gene regulatory sites in the upstream regulatory sequence, A and B, which are affected by the mutations. + indicates a normal site, and – indicates a mutant site that no longer binds its gene regulatory protein.

Transcription of the operon in different media

	Glucose only	Sucrose only	Glucose + sucrose
normal (A+ B+)	OFF	ON	OFF
mutant 1	OFF	OFF	OFF
mutant 2	OFF	ON	ON
mutant 3	OFF	OFF	OFF

(A) If mutant 1 has sites A–B+, which of these sites is regulated by sucrose and which by glucose?

(B) Give the state of the A and B sites in mutants 2 and 3.

(C) Which site is bound by a repressor, and which by an activator?

Initiation of Eucaryotic Gene Transcription Is a Complex Process (Pages 263–264)

8–16   Intermediate, multiple choice

Which one of the following is the main reason that a typical eucaryotic gene is able to respond to a far greater variety of regulatory signals than a typical procaryotic gene or operon?

    A. Eucaryotes have three types of RNA polymerase.

    B. Eucaryotic RNA polymerases require general transcription factors.

    C. The transcription of a eucaryotic gene can be influenced by proteins that bind far from the promoter.

    D. Eucaryotic genes are packaged into nucleosomes.

    E. The protein-coding regions of eucaryotic genes are longer than those of procaryotic genes.

Eucaryotic RNA Polymerase Requires General Transcription Factors (Pages 264–265)

8–17   Intermediate, multiple choice

Which of the following will prevent the release but not the assembly of the eucaryotic transcription initiation complex?

    A. Mutation of the TATA box.

    B. Absence of TBP.

    C. Absence of TFIIE.

    D. Absence of ATP.

    E. Absence of RNA polymerase.

Eucaryotic Gene Regulatory Proteins Control Gene Expression from a Distance (Pages 265–266)

8–18   Easy, multiple choice

How are most eucaryotic gene regulatory proteins able to affect transcription when their binding sites are far from the promoter?

    A. By binding to their binding site and sliding to the site of RNA polymerase assembly.

    B. By looping out the intervening DNA between their binding site and the promoter.

    C. By unwinding the DNA between their binding site and the promoter.

    D. By attracting RNA polymerase and modifying it before it can bind to the promoter.

Packing of Promoter DNA into Nucleosomes Can Affect Initiation of Transcription (Pages 266–267)

8–19   Intermediate, multiple choice

The expression of the *BRF1* gene in mice is normally quite low, but mutations in a gene called *BRF2* lead to increased expression of *BRF1*. You have a hunch that nucleosomes are involved in the regulation of *BRF1* expression and so you investigate the position of nucleosomes over the TATA box of *BRF1* in normal mice and in mice that either lack the *BRF2* protein (*BRF2–*) or part of histone H4 (*HHF–*) (histone H4 is encoded by the *HHF* gene). Your results are summarized below. + indicates a normal functional gene.

Mouse	Nucleosome positioning	Relative level of BRF1 mRNA
BRF2+ HHF+	specific pattern	1
BRF2– HHF+	random	100
BRF2+ HHF–	random	1
BRF2– HHF–	random	100

Which of the following conclusions CANNOT be drawn from your data?

A. *BRF2* is required for repression of *BRF1*.

B. *BRF2* is required for the specific pattern of nucleosome positions over the *BRF1* upstream region.

C. *HHF* is required for the specific pattern of nucleosome positioning over the *BRF1* upstream region.

D. The specific pattern of nucleosome positioning over the *BRF1* upstream region is required for *BRF1* repression.

E. The part of histone H4 missing in *HHF–* mice is not required to form nucleosomes.

Eucaryotic Genes Are Regulated by Combinations of Proteins (Pages 267–268)

8–20   Intermediate, matching/fill in blanks

The gene for a hormone necessary for insect development contains binding sites for three gene regulatory proteins called A, B, and C. Because the binding sites for A and B overlap, A and B cannot bind simultaneously. You make mutations in the binding sites for each of the proteins and measure hormone production in cells that contain equal amounts of the A, B, and C proteins. The results of your studies are summarized in Figure Q8–20.

In each of the following sentences, choose one of the phrases within square brackets to make the statement consistent with the above results.

A. A binds to its operator [more tightly / less tightly] than B binds to its operator.

B. A is a [stronger / weaker] activator of transcription than B.

C. C is able to prevent activation by [A only / B only / both A and B].

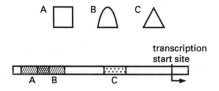

Binding site	Hormone production
A+ B+ C+	Off
A– B+ C+	Off
A+ B– C+	Off
A+ B+ C–	Low level
A– B– C+	Off
A+ B– C–	Low level
A– B+ C–	High level
A– B– C–	Off

Q8–20   – indicates a mutant site

The Expression of Different Genes Can Be Coordinated by a Single Protein (Pages 268–269)

8–21   Difficult, multiple choice

A virus produces a protein X that activates only a few of the virus's own genes when it infects cells. On examining the complete upstream gene regulatory sequences of three of the viral genes—V1, V2, and V3—that are activated by protein X you find that:

1. V1 and V2 contain binding sites for the zinc finger protein A only.

2. V3 contains a binding site for the homeodomain protein B only.

3. The only sequence that all three genes have in common is the TATA box.

In which one or more of the following ways could protein X activate transcription of V1, V2, and V3?

   A. Protein X binds nonspecifically to the DNA upstream of V1, V2, and V3 and activates transcription.
   B. Protein X binds to a repressor and prevents the repressor from binding upstream of V1, V2, and V3.
   C. Protein X activates transcription by binding to the TATA box.
   D. Protein X activates transcription by binding to the proteins A and B.
   E. Protein X represses transcription of the genes for proteins A and B.

Combinatorial Control Can Create Different Cell Types (Pages 269–271)

8–22   Easy, multiple choice

In principle, how many different cell types can an organism having four different types of gene regulatory proteins and thousands of genes create?

   A. Up to 4.
   B. Up to 8.
   C. Up to 12.
   D. Up to 16.
   E. Thousands.

Stable Patterns of Gene Expression Can Be Transmitted to Daughter Cells (Pages 271–273)

8–23   Difficult, multiple choice + short answer

A unicellular fungus exists in two states, a filamentous state and a yeastlike state, between which it switches every 10 to 20 cell divisions. A zinc-finger DNA-binding protein called PRY1 is continuously expressed in the yeastlike form. You have isolated a mutant strain of the fungus whose PRY1 protein unfolds at 37°C and thus is unable to bind DNA at this temperature. When yeastlike cells of this mutant strain are grown at 25°C and then switched to 37°C, the level of PRY1 mRNA drops sharply, but the cells remain yeastlike and switch to the filamentous state at the same rate as the normal strain does. But at 37°C, the filamentous state of the mutant fungus cannot switch to the yeastlike state. Which of the following statements about PRY1 are consistent with the behavior of this mutant strain? Explain your reasoning.

   A. PRY1 is a positive regulator of its own transcription.
   B. PRY1 is a negative regulator of its own transcription.

C. PRY1 is required to package the genes required for filamentous growth into a chromatin state in which they are repressed.

D. PRY1 is required to unpackage the genes required for yeastlike growth from a chromatin state in which they are repressed.

The Formation of an Entire Organ Can Be Triggered by a Single Gene Regulatory Protein (Pages 273–274)

8–24   Intermediate, multiple choice

In mammals, individuals that are XX are female, and individuals that are XY are male. It had long been known that a gene located on the Y chromosome was sufficient to induce the gonads to form testes, which is the main male-determining factor in development, and people began to search for the product of this gene, the so-called testis-determining factor (TDF). For several years, the TDF was incorrectly thought to be a zinc-finger protein encoded by a gene called *BoY*. Which of the following is the best evidence that *BoY* could NOT be the TDF?

A. Some XY individuals that develop into females have mutations in a different gene, *SRY*, but are normal at *BoY*.

B. *BoY* is not expressed in the adult male testes.

C. Expression of *BoY* in adult females does not masculinize them.

D. A few of the genes that are known to be expressed only in the testes have binding sites for the BoY protein in their upstream regulatory sequences, but most do not.

E. A gene encoding a protein whose amino acid sequence is very similar to that of the BoY protein is found on the X chromosome.

Answers

A8–1. A. In eucaryotic chromosomes, DNA is complexed with proteins to form <u>chromatin</u>.

B. The paternal and maternal copies of human Chromosome 1 are <u>homologous</u>.

C. The human X and Y chromosomes are <u>nonhomologous</u>.

D. Each mitotic chromosome is constricted at a site called the <u>centromere</u>.

E. Mitotic chromosomes are more <u>condensed</u> than <u>interphase</u> chromosomes.

A8–2. A.

A8–3. B.

A8–4. C.

A8–5. A. Nucleosomes are present in <u>eucaryotic</u> chromosomes, but not in <u>procaryotic</u> chromosomes.

B. A nucleosome contains two molecules each of histones <u>H2A and H2B</u>, as well as histones H3 and H4.

C. A nucleosome core particle contains a core of histone with DNA wrapped around it approximately <u>twice</u>.

D. Nucleosomes are aided in their formation by the high proportion of <u>basic</u> amino acids in histone proteins.

E. Nucleosome formation compacts DNA into approximately <u>one-third</u> of its original length.

A8–6. A. Interphase chromosomes contain both darkly staining <u>heterochromatin</u> and more lightly staining <u>euchromatin</u>.

B. A string of nucleosomes coils up with the help of <u>histone H1</u> to form the more compact structure of the <u>30-nm fiber</u>.

C. Genes that can be transcribed are thought to be contained in a relatively loosely packed type of euchromatin known as <u>active chromatin</u>.

D. Nucleosome core particles are separated from each other by a stretch of <u>linker</u> DNA.

E. After mitosis the mitotic chromatin <u>unfolds</u> to form the interphase chromosomes in the daughter cells.

A8–7. The state of chromatin packing can differ in all of the instances given. It differs most dramatically during the different stages of the cell cycle, but even within the same interphase chromosome, chromatin can be densely packed into heterochromatin or less tightly packed into euchromatin. In female mammals, one of the X chromosomes (which constitute a pair of homologous chromosomes) is entirely heterochromatic. Since different cell types in the same organism are expressing different genes, not all cells will have the same state of chromatin packing at any given time.

A8–8. D. Since the pattern of X-chromosome inactivation is set up randomly over several days of embryonic development, at a stage where the embryo has quite a few cells, it is unlikely that two cats will inactivate the same X chromosome in exactly the same set of cells. This makes C false also. The other statements are all false. If the mottled coloring is due to X-chromosome inactivation, then the coat-color gene involved must be on the X chromosome. Since males have only one X chromosome, which they receive from their mother, and do not undergo X-chromosome inactivation, males will not be mottled (A). Female calico cats receive an X chromosome from the father and from the mother, which may have different forms of the coat-color gene, and so are not necessarily the same color as their mother (B).

A8–9.　C and E. The variegation in color in the two types of peaches and of fuzziness in Old Man are hallmarks of position effects; the most likely explanation for the color of the two varieties is that the gene required for pinkness is in a region of heterochromatin, and is thus mainly repressed, leading to a mainly white skin. Since the Desert Peach is uniformly fuzzy, the gene required for fuzziness is likely to be expressed in all skin cells, thus D is not true. A chromosomal rearrangement that placed the fuzziness gene next to the region of hetero-chromatin that contains the pinkness gene would be likely to lead to its repression and could account for the appearance of Old Man. Mutations that inactivated gene regulatory proteins would be likely to lead to a uniform effect on fuzziness (thus A and B are unlikely).

A8–10.　C. Elevated levels of a protein that promotes heterochromatin formation would cause the genes in the region of heterochromatin to be more efficiently silenced, making Old Man more uniformly white and bald and Desert Peach more uniformly white. A protein that caused the entire chromosome to condense would cause Desert Peach to become uniformly white and bald. A repressor of the pinkness gene would not cause Old Man to become bald, and a repressor of the fuzziness gene would not cause Desert Peach and Old Man to become white.

A8–11.　E. The rRNA genes are all gathered in the nucleolus.

A8–12.　B. The major cause of differences between different cell types is in the differential expression of protein-coding genes transcribed by polymerase II, since these genes encode not only the speciality proteins characteristic of different cell types, but also the gene regulatory proteins required to maintain and control this pattern of expression. A is untrue, since all genes are replicated equally when cells divide. C is untrue, since expression of housekeeping genes do not differ much from cell to cell, as they mainly encode the proteins that are necessary for all cells to live. D is untrue, as the main stage at which gene expression is regulated is the initiation of transcription. E is untrue, as DNA is packed into nucleosomes in all eucaryotic cells.

A8–13.　A and B. Proteins that have the same DNA-binding motif share amino acid sequence similarity and have the same basic structure, but members of the same family do not necessarily bind to similar DNA sequences nor do family members necessarily control the same types of genes.

A8–14.　A. The genes of a bacterial operon are transcribed into a single mRNA.

B. Many bacterial promoters contain a region known as an operator to which a specific gene regulatory protein binds.

C. Genes in which transcription is prevented are said to be repressed.

D. Bacterial genes are regulated by small molecules such as tryptophan by the interaction of such molecules with allosteric DNA-binding proteins such as the tryptophan repressor.

E. Genes that are being constitutively expressed are being transcribed all the time.

A8–15.　(A) Site A is regulated by sucrose and site B by glucose. (B) Mutant 2 (A+ B–), mutant 3 (A– B–), or (A– B+). (C) Site A is bound by an activator and site B by a repressor.

A8–16.　C. The fact that, in eucaryotes, gene regulatory proteins can influence the initiation of transcription even when they are bound far away from the promoter means that there can be a very large number of gene regulatory sites affecting the same promoter. Thus the initiation of transcription can be influenced by a great variety and number of different signals, each of which may induce the binding of different gene regulatory proteins to these regulatory regions.

A8–17.　D. ATP is required for the phosphorylation of the RNA polymerase tail by the general transcription factor TFIIK, which allows the polymerase to be released. All of the other choices would prevent assembly of the initiation complex.

A8–18. B. Of the cases studied thus far, most eucaryotic gene regulatory proteins that act at a distance do so by looping out the intervening DNA while at the same time binding to proteins that form the initiation complex at the promoter.

A8–19. D. All the other conclusions can be drawn from the data. Since the *BRF2+ HHF2–* mutant does not exhibit the specific pattern of nucleosome positioning yet still has a low level of *BRF1* expression, and since the *BRF2–HHF2–* mutant has high levels of *BRF1* expression (indicating that *HHF2* is not required for *BRF1* expression), it would appear that repression of *BRF1* can take place in the absence of nucleosome positioning. Since nucleosomes are formed in all cases, the missing portion of histone H4 is not required for their formation.

A8–20. A. A binds to its operator <u>more tightly</u> than B binds to its operator.

B. A is a <u>weaker</u> activator of transcription than B.

C. C is able to prevent activation by <u>both A and B</u>.

When B is the only site present, the hormone is expressed strongly; therefore B is a strong activator. When A is the only site present, the hormone is expressed weakly; therefore, A is a weak activator. When A and B are present (but C is mutated), activation is weak; therefore A must be bound and not B (so A binds its DNA site better than B binds to its DNA site). Transcription is off whenever C is present, so C is able to repress activation by both A and B.

A8–21. D and E. If protein X bound nonspecifically to DNA, it would bind to any DNA, and would have no specific effect on any particular gene (thus A is unlikely). If there were a single repressor that bound upstream of all three genes, we would expect to have found a binding site common to all three genes, but there is none (thus B is unlikely). If protein X activated gene transcription by binding to the TATA box, it would be likely to activate almost all genes transcribed by RNA polymerase II, since they all contain TATA boxes (thus C is unlikely). Since we do not know whether proteins A and B are activators or repressors, it is possible that if protein X repressed the transcription of the genes encoding A and B, it would activate transcription of V1, V2, and V3.

A8–22. D. The type of cell is determined by the particular combination of gene regulatory proteins active within it. With four different proteins available, there is one possibility with no proteins at all and one with all four proteins. There are four possibilities with one protein each, six possible combinations of two different proteins and four possible combinations of three different proteins.

A8–23. A, C, and D are all consistent with the behavior of the mutant. Because a loss of PRY1 function (on switching to 37°C) causes an immediate drop in mRNA, PRY1 is a positive regulator of its own transcription (A). PRY1 is also required for the establishment, but not the maintenance of the yeastlike state, since filamentous forms cannot switch back to a yeastlike state at 37°C. Since a chromatin state can be inherited once established, both C and D would be consistent with the behavior of the mutant.

A8–24. A. XY individuals that develop as females presumably lack the testis-determining factor (TDF). If BoY is normal in these individuals, it would strongly suggest that BoY is not the TDF. Although expression of TDF is necessary for testes development, this does not mean that it must be expressed in adult males once the gonad has already formed. Similarly, even though TDF expression is sufficient to induce testis formation, once the structures are formed, TDF may not be able to exert any additional effect (thus B and C are not evidence). D is not compelling evidence against BoY being the TDF, since the TDF will not necessarily bind upstream of *all* of the genes whose expression it influences; some of the genes it regulates directly probably encode other gene regulatory proteins that bind to regulatory sites different from the TDF site. The presence of a protein similar to BoY on the X chromosome (E) is not necessarily evidence for or against BoY being the TDF.

9 Genetic Variation

Questions

GENETIC VARIATION IN BACTERIA (Pages 278–291)
The Rapid Rate of Bacterial Division Means That Mutation Will Occur Over a Short Time Period (Pages 279–280)

9–1   Easy, matching/fill in blanks (Requires information from section on pages 278–279)

For each of the following sentences, fill in the blanks with the correct word or phrase selected from the list below. Use each word or phrase only once.

   A. Bacteria are good subjects for studying genetic variation because a bacterial cell is _____ and thus carries _____ of each of its genes.

   B. A human liver cell carries _____ of the genome and thus is _____.

   C. Mutations are often detected by a change in the _____ of the organism.

   D. In a diploid organism, the effect of a mutation may be masked by the presence of an unmutated _____.

   E.   Bacteria divide asexually by _____.

phenotype; haploid; diploid; one copy; two copies; three copies; meiosis; allele; fission; mitosis.

9–2   Easy, multiple choice (Requires information from section on pages 285–288 and from Chapter 6)

Genetic variation is easily studied in bacteria because:

   A. bacteria are less susceptible to killing by mutagens than are higher organisms.

   B. the rapid rate of reproduction in bacteria leads to a higher rate of mistakes in DNA replication than in higher organisms.

   C. the rapid rate of reproduction in bacteria allows very large populations of bacteria to be studied.

   D. bacterial mismatch repair is not as efficient as eucaryotic mismatch repair.

   E. bacterial DNA does not undergo recombination.

Mutation in Bacteria Can Be Selected by a Change in Environmental Conditions (Pages 280–281)

9–3   Intermediate, multiple choice

Ura$^+$ bacteria can grow on medium that lacks the base uracil whereas mutant Ura$^-$ bacteria cannot. Ura$^-$ bacteria can, however, grow on medium that contains urabegone, a drug that kills Ura$^+$ cells. You inoculate a Ura$^+$ bacterium into media containing uracil and allow it to divide until there are 10^9 cells, which you then dilute and spread onto plates containing urabegone and uracil. You get 50 colonies in all. Which of the following statements are likely to be true?

   A. All of the cells in a given urabegone-resistant colony have the same mutation in a gene required for growth in the absence of uracil.

B. The cells in all 50 of the urabegone-resistant colonies all have the same mutation in a gene required for growth in the absence of uracil.

C. All of the Ura⁺ cells that did not grow on the urabegone plates were genetically identical.

D. If you inoculate a Ura⁺ bacterium and a Ura⁻ bacterium together into media containing uracil and grow them for a day, the Ura⁻ bacteria will compose only a small fraction of the population.

E. If you inoculate a Ura⁺ bacterium and a Ura⁻ bacterium together into media containing uracil and grow them for a day, the Ura⁻ bacteria will significantly outnumber the Ura⁺ population.

9–4 Easy, multiple choice (Requires information from Chapters 6 and 7)

The Ura⁻ bacteria described in Question 9–3 differ from the Ura⁺ bacteria in that:

A. the Ura⁺ bacteria do not use uracil in their cells.

B. the Ura⁻ bacteria lack an enzyme required to synthesize uracil.

C. the Ura⁻ bacteria lack an enzyme required to synthesize nucleotides from bases.

D. the Ura⁻ bacteria lack a transport protein required to take up uracil from the medium.

E. the Ura⁻ bacteria lack an enzyme that removes uracil nucleotides from DNA.

9–5 Intermediate, short answer

What are you likely to find if you inoculate a bacterium from a Ura⁻ colony described in Question 9–3 into medium containing uracil and then plate out 10^{12} of the resulting cells onto media lacking uracil? Explain your answer.

Bacterial Cells Can Acquire Genes from Other Bacteria (Pages 281–282)

9–6 Easy, matching/fill in blanks (Requires information from sections on pages 282–285 and pages 288–289)

For each of the following sentences which one, or more, of the three terms below could be used to complete it correctly. Write the appropriate numbers in the blank spaces.

A. _____ is mediated by bacteriophage.

B. _____ involves transfer of DNA between living cells.

C. _____ involves direct uptake of DNA from the environment.

D. _____ results in a higher frequency of genetic change than can be accounted for by mutation due to replication errors.

E. _____ does not require cell-to-cell contact.

1. Bacterial mating; 2. Transformation; 3. Transduction.

Bacterial Genes Can Be Transferred by a Process Called Bacterial Mating (Pages 282–284)

9–7   Easy, multiple choice

A normal wild-type *E. coli* can grow on medium lacking both methionine and leucine. A mutant Met⁻ strain of *E. coli*, which also contains an F plasmid, cannot grow on medium lacking methionine. A Leu⁻ strain, which lacks an F plasmid, cannot grow on medium lacking leucine. After mixing Met⁻ and Leu⁻ bacteria together for a few hours you transfer the culture to medium lacking both methionine and leucine and obtain immediate bacterial growth in the new medium. Which of the following is most likely to have occurred?

   A. The mutant *met* gene of the Met⁻ strain has been restored to normal function by a further mutation in it.

   B. A functional *met* gene has been transferred by conjugation from the Leu⁻ strain to the Met⁻ strain.

   C. The mutant genes in both strains of bacteria have been restored to normal function by further mutations.

   D. A *leu* gene has been transferred by conjugation from the Met⁻ strain to the Leu⁻ strain.

   E. A *met* gene has been transferred by conjugation from the Met⁻ strain to the Leu⁻ strain.

9–8   Intermediate, short answer

You have two strains of bacteria, one of which is F⁺ and has the phenotype Leu⁺ Met⁺ His⁺ Arg⁺ Lys⁺ (meaning it can make its own leucine, methionine, histidine, arginine, and lysine and thus grow in their absence) and one of which is Leu⁻ Met⁻ His⁻ Arg⁻ Lys⁻, and thus requires all these amino acids to be supplied. You allow the two strains to mate, plate out the bacteria that have arisen from the mating and then determine the phenotype of the bacteria in the individual colonies that result by examining their ability to grow on various media. Your results are summarized in Figure Q9–8.

Sketch a map of the bacterial chromosome indicating (1) the relative positions of the genes responsible for allowing growth in the absence of each amino acid and (2) the position of the integrated F plasmid.

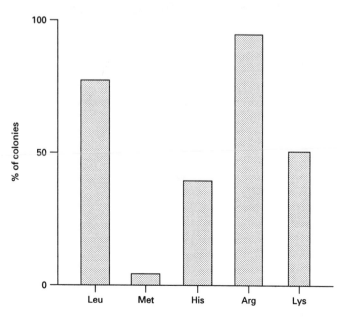

Q9–8

Some Bacteria Can Take Up DNA from Their Surroundings (Pages 284–285)

9–9 Easy, multiple choice

Which of the following are advantages of using transformation rather than bacterial mating as a means of introducing new genes into bacteria?

    A. Transformation provides a way of introducing DNA from organisms other than bacteria.

    B. Transformation enables bacteria to acquire DNA from other bacteria without killing the donor bacteria.

    C. Transformation removes the need for homologous recombination.

    D. Transformation provides a mechanism for protecting the DNA during transfer.

Gene Exchange Occurs by Homologous Recombination Between Two DNA Molecules of Similar Nucleotide Sequence (Pages 285–288)

9–10 Difficult, short answer (Requires information from Chapter 6)

A bacterial enzyme required for sucrose metabolism is encoded by the *suc* gene, which is depicted schematically in Figure Q9–10A. The numbers represent the number of nucleotides along the gene. Flanking DNA sequences not involved in expression of the *suc* gene are represented by the thick black line.

You have constructed four plasmids containing various pieces of the *suc* gene. Each plasmid also contains an ampicillin-resistance marker gene (amp^r) but no origin of replication. The thin line represents plasmid sequences that are not homologous to any DNA in the bacterial genome. Which of the plasmids depicted in Figure Q9–10B will give you Suc⁻ Ampʳ colonies (in which the bacteria cannot utilize sucrose) when transformed into a Suc⁺ Ampˢ (ampicillin-sensitive) strain? Explain your reasoning.

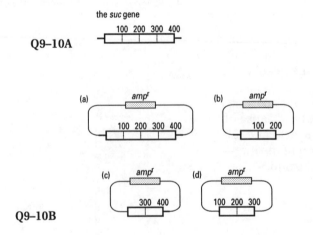

Genes Can Be Transferred Between Bacteria by Bacterial Viruses (Pages 288–289)

9–11   Intermediate, short answer

The lambdalike virus Q9 is capable of growing as a prophage or multiplying by the lytic pathway. Like lambda, Q9 is capable of *specialized transduction*, or transduction of genes caused by the inaccurate excision of an integrated prophage. However, when Q9 grows lytically, it digests the host genome before viral assembly is complete and sometimes packages pieces of bacterial DNA into the viral coat instead of phage DNA; this is called *generalized transduction*. Indicate how each of the following mutations in the cell in which the transducing phage is formed would affect the formation of specialized and generalized transducing phages.

1. A mutation of the Q9 integrase that destroys its function.

2. A mutation in Q9 that prevents circularization of the virus.

3. A mutation that prevents homologous recombination.

4. A mutation in the site in the host genome recognized by integrase.

5. A mutation in the coat protein of the virus.

9–12   Intermediate, short answer

Which of mutations 1–4 listed in Question 9–11 would be most likely to affect transduction if it were in the recipient cell that becomes infected by a generalized transducing phage? How would transduction be affected?

Transposable Elements Create Genetic Diversity (Pages 289–291)

9–13   Intermediate, multiple choice (Requires information from section on pages 296–297)

What is the most common result of the insertion of a transposable element into a bacterial genome?

    A. Change in the level of expression of nearby genes.

    B. Induction of homologous recombination of the genome.

    C. Inactivation of a gene.

    D. Acquisition of drug resistance by the bacterium.

    E. Movement of genes from one part of the bacterial chromosome to another.

SOURCES OF GENETIC CHANGE IN EUCARYOTIC GENOMES (Pages 291–304)
Random DNA Duplications Create Families of Related Genes (Pages 292–293)

9–14   Easy, multiple choice

Members of the same gene family share sequence similarity:

    A. in all of the exons but none of the introns.

    B. in all of the introns but none of the exons.

C. over the entire sequence of the gene.

D. in some but not all of the exons.

E. in the regulatory regions only.

Genes Encoding New Proteins Can Be Created by the Recombination of Exons (Pages 293–294)

9–15   Intermediate, multiple choice

Which of the following would contribute most to successful exon shuffling?

A. Shorter introns.

B. A haploid genome.

C. Exons that code for more than one protein domain.

D. Introns that contain regions of similarity to one another.

E. Inability of short stretches of amino acids to fold into discrete functional units.

A Large Part of the DNA of Multicellular Eucaryotes Consists of Repeated, Noncoding Sequences (Pages 294–295)

9–16   Easy, multiple choice

Satellite DNA:

A. makes up about one-third of the human genome.

B. has important gene regulatory functions in the genome.

C. is evenly distributed throughout the genome.

D. is derived from two types of transposable elements.

E. is composed of clusters of repeated short sequences.

About 10% of the Human Genome Consists of Two Families of Transposable Sequences (Pages 295–296)

9–17   Easy, multiple choice

Retrotransposons:

A. are found only in eucaryotes.

B. can move by either the cut-and-paste mechanism or by a mechanism requiring an RNA intermediate.

C. include the transposable elements LINE-1, Alu, and Tn10.

D. can move only if they encode a reverse transcriptase.

The Evolution of Genomes Has Been Accelerated by Transposable Elements (Pages 296–297)

9–18   Difficult, short answer

After looking at Figure Q9–18, describe two quite different mechanisms by which the gene that encodes protein C could have arisen from the genes that encode proteins A and B.

Q9–18

Protein A   NH₂—[X]—(Z)—COOH

Protein B   NH₂—(Y)—(Y)—(Y)—COOH

Protein C   NH₂—[X]—(Y)—(Z)—COOH

Viruses Are Fully Mobile Genetic Elements That Can Escape from Cells (Pages 297–300)

9–19   Easy, multiple choice (Requires information from section on pages 300–302)

All viruses:

    A. have single-stranded genomes.

    B. lyse the cells they infect.

    C. encode all of the enzymes needed to replicate themselves.

    D. contain both nucleic acid and protein.

    E. have the same size genomes.

9–20   Easy, multiple choice

Which of the following molecules required for replication must be encoded by an RNA virus such as poliovirus?

    A. RNA polymerase.

    B. RNA replicase.

    C. Primase.

    D. Transfer RNA.

    E. Ribosomal RNA.

9–21   Intermediate, art labeling

Having isolated a virus, you could start to identify it by finding out what type of nucleic acid it contains. Fill in the blank boxes in Figure Q9–21 to provide a simple key that sets out the possible alternatives.

Q9–21

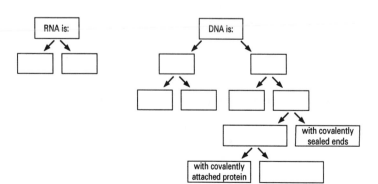

9–22   Easy, short answer

What two distinguishing features of a virus that would be visible in the electron microscope would help you to identify it?

Retroviruses Reverse the Normal Flow of Genetic Information (Pages 300–302)

9–23   Easy, multiple choice (Requires information from sections on pages 295–297)

The enzymes reverse transcriptase and DNA polymerase both synthesize DNA. Which of the following statements about them are true?

    A. Reverse transcriptase uses only an RNA template, DNA polymerase uses only a DNA template.

    B. Reverse transcriptase can use either an RNA or a DNA template, DNA polymerase uses only a DNA template.

    C. DNA polymerase is used only by cells, reverse transcriptase is used only by viruses.

    D. DNA polymerase uses deoxynucleotides, reverse transcriptase uses ribonucleotides.

9–24   Easy, multiple choice

Retroviruses:

    A. have double-stranded circular RNA genomes.

    B. do not require RNA polymerase to complete their replication cycle.

    C. do not lyse the host cell in order to escape into the environment.

    D. encode integrases that catalyze the integration of the viral genome into a specific site in the host genome.

    E. are non-enveloped viruses.

9–25   Intermediate, short answer

Why would it not be possible to eradicate a retrovirus that has integrated into the genome of a host cell by treatment of the infected cells with reverse transcriptase inhibitors?

9–26   Intermediate, short answer

Why do retroviruses need to package reverse transcriptase molecules into their virus particles even though they carry the gene for reverse transcriptase in their genomes?

Retroviruses That Have Picked Up Host Genes Can Make Cells Cancerous (Pages 302–304)

9–27   Intermediate, multiple choice

Which of the following statements regarding RNA tumor viruses are true?

A. Only retroviruses that carry the *src* gene are capable of producing sarcomas.

B. Most proto-oncogenes encode regulators of transcription.

C. Retroviruses can convert a proto-oncogene to an oncogene either because the proto-oncogene becomes mutated or because its pattern of expression becomes changed.

D. Retroviruses tend to pick up oncogenes because they help the viruses spread from host to host.

E. Because retroviruses integrate into the host genome, they produce cancer in all of the host's offspring.

SEXUAL REPRODUCTION AND THE REASSORTMENT OF GENES (Pages 304–309)
Sexual Reproduction Gives a Competitive Advantage to Organisms in an Unpredictably Variable Environment (Pages 304–305)

9–28   Intermediate, multiple choice

Sexual reproduction:

A. introduces new genes into a population.

B. enables a population to get rid of deleterious alleles because offspring that inherit these alleles die.

C. selects for those genes that make individuals better adapted to the prevailing conditions.

D. produces offspring that are different from either parent and from each other.

9–29   Intermediate, short answer

Why is sexual reproduction more beneficial to a species living in an unpredictable environment than to one living in a constant environment?

Sexual Reproduction Involves Both Diploid and Haploid Cells (Pages 305–306)

9–30   Easy, multiple choice (Requires information from section on pages 306–307)

Which of the following statements are true?

A. Diploid organisms reproduce only sexually.

B. All sexually reproducing organisms have at least two copies of each gene at some stage in their life cycle.

C. Gametes have only one chromosome.

D. Another name for the unfertilized egg cell is the zygote.

E. Only two haploid cells are produced from one diploid cell by meiosis.

Meiosis Generates Haploid Cells from Diploid Cells (Pages 306–307)

9–31   Intermediate, multiple choice

If meiosis in a diploid germ cell precursor became temporarily arrested just before completion of the first meiotic cell division and a transposon hopped into a gene during this period of arrest, how many of the gametes produced from this precursor cell would be mutant?

    A. 0/4.

    B. 1/4.

    C. 2/4.

    D. 3/4.

    E. 4/4.

Meiosis Generates Enormous Genetic Variation (Pages 307–309)

9–32   Intermediate, short answer

(A) Assuming recombination does not occur, what is the maximum number of genetically different types of sperm cells that could be produced in an animal with three pairs of homologous chromosomes in its diploid cells?

(B) How many genetically different offspring could, in theory, be produced by one pair of unrelated individuals of the above species?

9–33   Difficult, multiple choice

A certain type of worm has two genders: males that produce sperm and hermaphrodites that produce both sperm and eggs. The diploid adult has four homologous pairs of chromosomes that undergo very little recombination. Given a choice, the hermaphrodites prefer to mate with males, but just to annoy the worm, you pluck a hermaphrodite out of the wild and fertilize its eggs with its own sperm. Assuming all the resulting offspring are viable, what fraction do you expect to be genetically identical to the parent worm? Assume that each chromosome in the original hermaphrodite is genetically distinct from its homologue.

    A. All.

    B. None.

    C. Half.

    D. 1/16.

    E. 1/256.

Answers

A9–1.   A. Bacteria are good subjects for studying genetic variation because a bacterial cell is <u>haploid</u> and thus carries <u>one copy</u> of each of its genes.

B. A human liver cell carries <u>two copies</u> of the genome and thus is <u>diploid</u>.

C. Mutations are often detected by a change in the <u>phenotype</u> of the organism.

D. In a diploid organism, the effect of a mutation may be masked by the presence of an unmutated <u>allele</u>.

E. Bacteria divide asexually by <u>fission</u>.

A9–2.   C. In order to study genetic variation, we need to be able to look at very large populations, and we need to be able to see how the population changes with each generation. This is easy with bacteria because it is easy to maintain large populations cheaply and conveniently and because the time between each generation is short. Moreover, a large population of bacteria can be obtained from a single cell in less than a day. A is untrue; if anything, bacteria tend to be more easily killed by mutation than are higher organisms because they have only one copy of each gene. B and D are untrue; bacterial repair and replication are comparable in efficiency to their eucaryotic counterparts. E is untrue.

A9–3.   A. Since all of the cells in a given colony arise from a single cell and all cells that grow on urabegone plates are Ura⁻, it is highly likely that all of the cells in a given colony have the same mutation in the gene that allows growth in the absence of uracil. These experiments do not tell you whether the 50 Ura⁻ cells that gave rise to the 50 urabegone-resistant colonies were produced independently (B). All 50 colonies could be formed from 50 descendants of a single mutant cell; it is also possible that more than one cell has sustained a (different) mutation in the *ura* gene. Nor does it tell you anything about the genetic constitution of the Ura⁺ bacteria; in a population of 10^9 bacteria you could expect mutations in many different genes to have occurred (C). In media containing uracil, there is no significant selective advantage to being Ura⁺, so there is no obvious reason for the Ura⁻ gene either to be lost from the population (D) or to take over the population (E).

A9–4.   B. A must be untrue as all living organisms use uracil (U) as a base in RNA. If C or D were true, the Ura⁻ bacteria would not live even if supplied with uracil in the medium. E would increase the mutation rate but not the ability of bacteria to grow in the absence of uracil.

A9–5.   You would probably get at least a few colonies. From the results of the experiment described in Question 9–3 you can deduce that the frequency of getting a spontaneous mutation is likely to be about 50 in 10^9, so chances are, if you grow 10^{12} Ura⁻ cells from a single cell, at least one will mutate back to being Ura⁺ and thus be able to grow on medium lacking uracil.

A9–6.   A, 3; B, 1; C, 2; D, 1, 2, and 3; E, 2 and 3.

A9–7.   D. It is unlikely that the specific mutations required (A and C) would have arisen at such a high frequency that they would be found after a few hours' incubation. The appearance of Met⁺Leu⁺ bacteria over a short time period indicates that DNA has been transferred by conjugation. Since the Met⁻ strain is the one carrying the F⁺ plasmid, the direction of transfer will have been (at least initially) from Met⁻ bacteria (not from Leu⁻ bacteria, as in B), and a functional *leu* gene will have been transferred. Transfer of the *met* gene from Met⁻ bacteria (E) would have not enabled the recipient bacteria to grow on leucine.

A9–8.   Figure A9–8. The genes closest to the
        integrated F plasmid will be transferred with
        the highest frequency. Hence, *arg*
        is closest to the F plasmid, then *leu,*
        then *lys,* then *his,* then *met.*

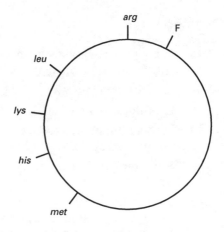

                      A9–8

A9–9.   A. Since transformation involves uptake of DNA from the environment, transformation allows
        cells to use DNA from any organism. Since the DNA must be naked in order for cells to be
        transformed, the donor cell must be lysed (i.e., killed) to release the DNA (B); in addition, while
        the DNA is naked, it is vulnerable to nucleases and other damaging agents in the environment
        (D). When linear DNA or DNA lacking a replication origin is transformed into a cell, the only
        way the bacterium can maintain the DNA is by inserting it into its genome by homologous
        recombination (C).

A9–10. D. The plasmids have no origin of replica-
        tion, so in order to be maintained, they
        must integrate into the genome via the
        homology to the *suc* gene. All of the plas-
        mids except (d) will yield at least one full-
        length *suc* gene after integration, as shown
        for (a) and (b), for example, in Figure A9–10
        (recombination at only one of the possible
        sites is shown in each case). Plasmid (d)
        yields two incomplete fragments of the *suc*
        gene after integration and hence makes the
        cells Suc⁻.

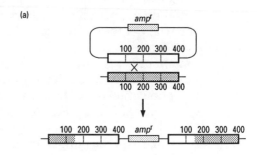

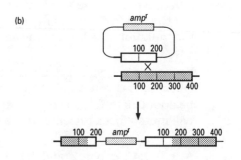

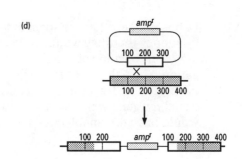

                      A9–10

A9–11. 1 and 4 would decrease specialized transduction only; 2 and 5 would decrease both; 3 would affect neither. Integrase and the attachment site in the host genome are both required only for formation of the prophage; therefore, mutations in either will affect only the formation of a specialized transducing phage. Homologous recombination is not required for the formation of either type of phage. Circularization of the virus is required for both lytic and prophage pathways and therefore is required for formation of both specialized and generalized transducing phages. Likewise, in order for any transducing phage to be released, new phage particles must be made, so a mutation in the coat protein will affect both types of transduction.

A9–12. 3. If the host bacterium were unable to carry out homologous recombination, it would not be able to incorporate the incoming bacterial DNA contained in the generalized transduction particle into its chromosome, and thus could not pass it on to its progeny. None of the other mutations would have any effect, as the incoming generalized transducing particle does not carry a phage genome.

A9–13. C. In bacteria, genes are close together and have no introns; therefore it is highly likely that when a transposon hops into the genome, it will insert into a gene and inactivate it. In order for a transposon to change the level of gene expression (A), it must insert close to a gene without disrupting it. Many transposons do not carry drug-resistance genes and therefore do not cause drug resistance when they insert into the genome (D). Induction of homologous recombination (B) or movement of a gene sandwiched between two transposons (E) requires that two independent transposition events involving the same or similar transposons occur.

A9–14. C. Gene families often arise by the duplication of a whole gene by misalignment and recombination. Hence members of the same gene family tend to share similarities over the entire sequence of the gene, in both the introns and exons; the DNA sequences of the exons, however, are typically more similar to one another than they are to the introns.

A9–15. D. Exon shuffling is facilitated by long introns (thus A is incorrect) between short exons that each code for one protein domain (thus C is incorrect). Since exon shuffling can occur via recombination between introns, introns with regions of similarity to one another will facilitate shuffling. A haploid genome will probably be *less* prone to exon shuffling than a diploid genome (thus B is incorrect) because having two copies of each gene allows an organism to keep a backup copy of the gene while it shuffles the other. Exon shuffling is possible only because many proteins are modular, composed of short, folded domains that have discrete functional properties (thus E is incorrect).

A9–16. E. Satellite DNA is clusters of repeated short sequences found near the telomeres and centromeres of eucaryotic chromosomes (thus C is incorrect). It makes up approximately 10% of the human genome (A is incorrect) and is much less complex than the transposable elements that also litter the human genome (D is incorrect). It has at present no known function (B is incorrect).

A9–17. A. Retrotransposons move only through an RNA intermediate (so B is incorrect). They include LINE-1 and Alu sequences (but not Tn10, which is a bacterial transposon) and do not necessarily need to provide their own reverse transcriptase so long as a similar transposon that does encode reverse transcriptase is also present in the genome (so C and D are incorrect).

A9–18. Figure A9–18. (1) A piece of DNA that encodes domain Y could have been transposed into an intron between the exons for X and Z. (2) Two recombination events could have occurred between the chromosomes that encode proteins A and B.

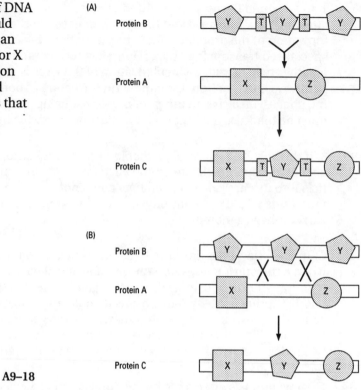

A9–18

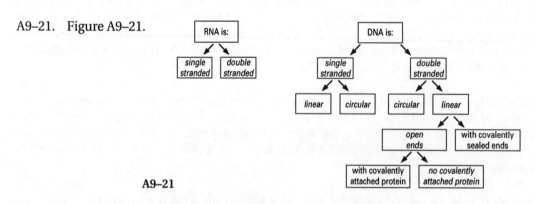

A9–21

A9–19. D. All viruses contain both protein and nucleic acid. Viruses can have either double- or single-stranded genomes (A). Not all viruses lyse the cells they infect (B); for example, some bud out of the cell without killing it. Viruses can have as few as three genes or more than a hundred (E). No virus is able to replicate in the absence of a host cell (C).

A9–20. B.

A9–21. Figure A9–21.

A9–22. Its shape and whether or not it had an outer lipid envelope.

A9–23. B.

A9–24. C. Retroviruses infect only eucaryotic cells and reproduce by budding. Retroviral integrase catalyzes the integration of the viral genome into random sites in the host genome.

A9–25.   Once the virus has integrated into the genome, it has no further need for reverse transcriptase. Therefore, an inhibitor of reverse transcriptase may be able to block infection of other cells by viruses that bud off the infected cell but will not be able to eradicate the integrated virus.

A9–26.   Retroviruses carry their own reverse transcriptase with them, as they must produce a double-stranded DNA copy of their genome before their genes can be transcribed and expressed.

A9–27.   C. Many oncogenes other than *src* cause sarcomas (A). Proto-oncogenes encode many types of proteins that affect regulation of cell division or differentiation (B). Some proto-oncogenes are transcriptional regulators, but many others are receptors or other components of cell signaling pathways. In general oncogenes do not help the retrovirus (D). Retroviruses produce cancer in the offspring of the original host only if the infection takes place in germ-cell precursors (E).

A9–28.   D. Sexual reproduction shuffles the existing genes in a population but cannot actually introduce new genes (A). Because sexual organisms are diploid, individuals that inherit a deleterious allele do not necessarily die unless one normal copy of the gene is insufficient to support life (B). Sexual reproduction itself does not select for genes that make individuals better adapted. Indeed, by shuffling alleles randomly in each generation it is just as likely to break up desirable combinations of alleles (C).

A9–29.   The real benefit in sexual reproduction seems to be that parents produce children that are genetically unlike either parent and that are not genetically identical to each other. This provides more variation in the population than asexual reproduction could provide, and is an advantage if the environment is variable, since combination of the parents' characteristics, however well adapted to the prevailing conditions, may or may not be the best in a new situation.

A9–30.   B. The sexual life cycle involves an alternation of diploid and haploid phases; the diploid phase gives rise to a haploid phase (in multicellular animals this is limited to the gametes only), which produces gametes that combine by the process of fertilization to produce a new diploid. Some diploid organisms (e.g., many plants) are capable of asexual reproduction (A). Gametes have only one member of each pair of homologous chromosomes, but since most organisms have more than one pair of homologous chromosomes, most gametes have more than one chromosome (C). D is untrue: the zygote is formed by fusion of the egg and sperm. E is untrue: four haploid cells are produced from one diploid cell by meiosis.

A9–31.   B. Prior to the first meiotic division the chromosomes have all replicated, and so there are now four copies of each type of chromosome (two copies of the maternal homologue and two copies of the paternal homologue), each of which will go to a different gamete. Since replication has already occurred, if a transposon hops into a gene during the arrest period, only one chromosome will be affected and 1 out of 4 gametes will be mutant.

A9–32.   (A) If the paternal and maternal chromosomes of each chromosome pair are not genetically identical, the animal can produce $2^3 = 8$ genetically different sperm. (B) The male can produce 8 genetically different sperm and the female can produce 8 genetically different eggs, so the maximum number of genetically different offspring is $8 \times 8 = 64$.

A9–33. D. The parent is most likely heterozygous for each chromosome, and so can produce $2^4 = 16$ types of eggs and $2^4 = 16$ types of sperm. Any of the eggs produced will be able to give rise to an adult that is identical to the parent, but in order to do so, it must be fertilized by the right type of sperm. For each type of egg, only one of the 16 possible sperm will produce a diploid that is identical to the parent. Therefore, 1 out of 16 of the offspring should be identical to the parent. In other words, a sexually reproducing organism with several chromosomes has a relatively high probability of producing genetically distinct offspring even when the parent mates to itself.

10 DNA Technology

Questions

HOW DNA MOLECULES ARE ANALYZED (Pages 315–320)

10–1 Easy, multiple choice (Requires information from Chapter 2)

It is easiest to determine the sequence of subunits in:

    A. proteins.

    B. a straight-chain polysaccharide composed of different monomers.

    C. a branched-chain polysaccharide composed of different monomers.

    D. DNA.

    E. RNA.

Restriction Nucleases Cut DNA Molecules at Specific Sites (Pages 315–317)

10–2 Easy, matching/fill in blanks (Requires information from sections on pages 315–320)

For each of the following sentences, fill in the blanks with the correct word or phrase selected from the list below. Use each word or phrase only once.

    A. A nuclease hydrolyzes the _____ bonds in a nucleic acid.

    B. Nucleases that cut DNA only at specific short sequences are known as _____.

    C. DNA composed of sequences from different sources is known as _____.

    D. _____ can create organisms with combinations of genes that probably have never occurred naturally.

    E. Millions of copies of a DNA sequence can be made entirely *in vitro* by _____.

hydrogen; restriction nucleases; phosphodiester; endonucleases; ribonucleases; recombinant DNA; genetic engineering; exonucleases; the polymerase chain reaction; nucleic acid hybridization.

10–3 Easy, multiple choice

The sites at which restriction enzymes cut:

    A. are found only in bacterial DNA.

    B. are usually palindromic.

    C. are all found at the same frequency in DNA.

    D. are the same number of nucleotides long.

    E. produce staggered ends when cut.

10–4   Intermediate, multiple choice

You have purified DNA from your recently deceased goldfish. Which of the following restriction nucleases would you use if you wanted to end up with DNA fragments of average size around 70 kb after complete digestion of the DNA? The recognition sequence for each enzyme is as indicated in the right-hand column. "N" indicates that any nucleotide may be in this position and the enzyme will still cleave the site.

    A. Sau 3aI   GATC
    B. Bam HI   GGATCC
    C. Bsa JI   CCNNGG
    D. Not I   GCGGCCGC
    E. Xxx I   GAAGGATCCTTC

Gel Electrophoresis Separates DNA Fragments of Different Sizes (Pages 317–320)

10–5   Intermediate, short answer

You have accidentally torn the labels off two tubes each containing a different plasmid and now do not know which plasmid is in which tube. Fortunately, you have restriction maps for both plasmids, shown in Figure Q10–5.

You have the opportunity to test just one sample from one of your tubes. You have equipment for agarose gel electrophoresis, a standard set of DNA size markers, and the necessary restriction enzymes.

(A) Outline briefly the experiment you would do to determine which plasmid is in which tube.

(B) Which restriction enzyme or combination of restriction enzymes would you use in this experiment?

Q10–5

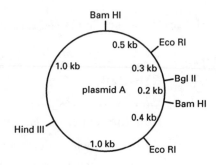

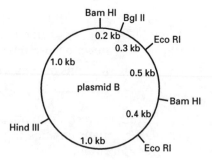

The Nucleotide Sequence of DNA Fragments Can Be Determined (Page 320)

10–6   Easy, short answer

You have sequenced a short piece of DNA and produced the gel shown in Figure Q10–6. What is the sequence of the DNA (starting from the 5' end).

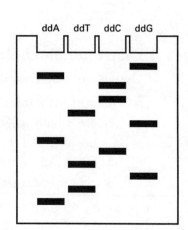

Q10–6

10–7 Difficult, multiple choice + short answer

You have sequenced a fragment of DNA and produced the gel shown in Figure Q10–7.

Near the top of the gel, there is a section where there are bands in all four lanes (indicated by the arrow). Which of the following mishaps would account for this phenomenon? Explain your answer.

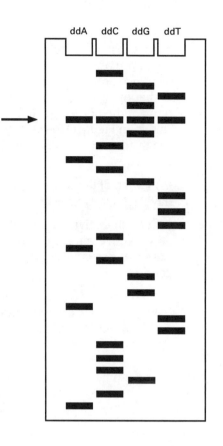

A. You mistakenly added all four dideoxynucleotides to one of the reactions.

B. You forgot to add deoxynucleotides to the reactions.

C. You forgot to add dideoxynucleotides to each of the reactions.

D. A fraction of the DNA you are sequencing was cut by a restriction nuclease.

E. Your primer hybridizes to more than one area of the fragment of DNA you are sequencing.

Q10–7

NUCLEIC ACID HYBRIDIZATION (Pages 320–324)

10–8 Easy, matching/fill in blanks (Requires information from sections on pages 320–324)

For each of the following sentences, fill in the blanks with the correct word or phrase selected from the list below. Use each word or phrase only once.

A. The technique of _____ hybridization can be used to detect a specific DNA sequence on a chromosome.

B. Northern blotting detects a specific sequence in _____.

C. Southern blotting detects a specific sequence in _____.

D. A short single-stranded DNA is an _____.

E. A piece of DNA used to detect a specific sequence in a nucleic acid by hybridization is known as a _____.

RNA; DNA; oligonucleotide; the polymerase chain reaction; *in situ*; *in vivo*; vector; probe.

DNA Hybridization Facilitates the Prenatal Diagnosis of Genetic Diseases (Pages 321–323)

10–9 Difficult, short answer

Assume that defects in a hypothetical gene, X, have been linked to antisocial behavior. Two copies of a defective gene X predispose a child to bad behavior from childhood, while a single copy of the gene seems to produce no symptoms until adulthood. Since the effects of the gene can be counteracted if treatment is started early enough, a program of voluntary genetic testing is being carried out with delinquent prospective parents. Charles S. and Caril Ann F. have been arrested on charges of robbery and assault, and Caril Ann is pregnant with Charles's child. You obtain DNA samples from Charles, Caril Ann, and the fetus, and on each you perform two Southern blots using Not I to cleave the DNA and two oligonucleotide probes, A and B, that hybridize to different parts of the normal gene X, as shown in Figure Q10–9A. You get the results shown in Figure Q10–9B.

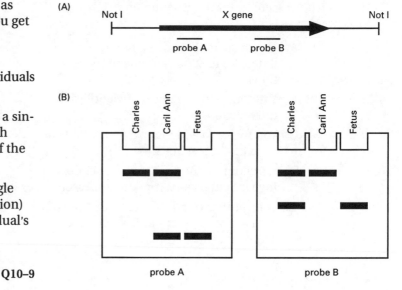

(A) Which of the three individuals have defects in gene X?

(B) Which individuals have a single defective gene and which have two defective copies of the gene?

(C) Indicate the nature (single base-pair mutation or deletion) and location of each individual's defects on gene X.

Q10–9

In Situ Hybridization Locates Nucleic Acid Sequences in Cells or on Chromosomes (Pages 323–324)

10–10   Intermediate, short answer

Figure Q10–10A shows a restriction map of a piece of DNA containing your favorite gene. The arrow indicates the position and orientation of the gene in the DNA. In part B of the figure are enlargements showing the portions of the DNA whose sequences have been used to make oligonucleotide probes A, B, C, and D. Which of the oligonucleotides can be used to detect the gene in each of the following:

1. A Southern blot of genomic DNA cut with Hind III.

2. A Southern blot of genomic DNA cut with Bgl II.

3. A Northern blot.

Q10–10

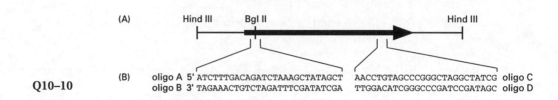

10–11   Easy, multiple choice

In situ hybridization can be used to determine:

   A. the sequence of a cloned gene.
   B. the distribution of proteins in tissues.
   C. the position of a cloned fragment of DNA on a plasmid.
   D. the size of a gene.
   E. the distribution of a given type of mRNA in different tissues.

DNA CLONING (Pages 324–335)

10–12   Easy, matching/fill in blanks (Requires information from sections on pages 324–335)

For each of the following sentences, fill in the blanks with the correct word or phrase selected from the list below. Use each word or phrase only once.

   A. Two fragments of DNA can be joined together by _____.
   B. Restriction enzymes that cut DNA straight across the double helix produce fragments of DNA with _____.
   C. A fragment of DNA is inserted into a _____ in order to be cloned in bacteria.
   D. A _____ library contains a collection of DNA clones derived from mRNAs.
   E. A _____ library contains a collection of DNA clones derived from chromosomal DNA.

staggered ends; DNA ligase; genomic; DNA polymerase; blunt ends; RNA; cDNA; vector; probe.

DNA Ligase Joins DNA Fragments Together to Produce a Recombinant DNA Molecule (Pages 325–326)

10–13   Intermediate, short answer (Requires information from Chapter 6)

Figure Q10–13 shows the recognition sequences and sites of cleavage for the restriction enzymes Sal I, Xho I, Sma I, and Pst I, and a plasmid with the sites of cleavage for these enzymes marked.

(A) After which of the following treatments 1–5 can the plasmid shown in Figure Q10–13 be recircularized simply by treating with DNA ligase? Assume that after treatment, any small pieces of DNA are removed and it is the larger portion of plasmid only that you are trying to recircularize.

After digestion with:
   1. Sal I alone.
   2. Sal I and Xho I.
   3. Sal I and Pst I.
   4. Sal I and Sma I.
   5. Sma I and Pst I.

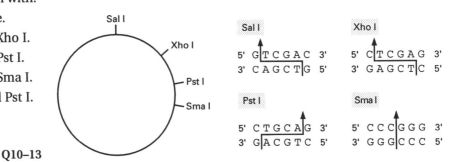

Q10–13

(B) In which of the cases 1–5 can the plasmid be recircularized by adding DNA ligase after the cut DNA has been treated with DNA polymerase in a mixture containing the four deoxynucleotides? Again assume that you are trying to recircularize the larger portion of plasmid.

Bacterial Plasmids Can Be Used to Clone DNA (Pages 326–327)

10–14   Easy/Intermediate, short answer

Name three features that a cloning vector for use in bacteria must contain. Explain your answers.

Human Genes Are Isolated by DNA Cloning (Pages 327–329)

10–15   Intermediate, multiple choice

Which of the following statements about genomic DNA libraries are false?

A. The larger the size of the fragments used to make the library, the fewer colonies you will have to examine to find a clone that hybridizes to your probe

B. The larger the size of the fragments used to make the library, the more difficult it will be to find your gene of interest once you have identified a clone that hybridizes to your probe.

C. The larger the genome of the organism from which a library is derived, the larger the fragments inserted into the vector will tend to be.

D. The smaller the gene you are seeking, the more likely it is that the gene will be found on a single clone.

E. The shorter the oligonucleotide used to probe the library, the greater the number of colonies to which the probe will hybridize.

10–16   Intermediate, multiple choice

A DNA library has been constructed by purifying chromosomal DNA from mice, cutting the DNA with the restriction enzyme Not I, and inserting the fragments into the Not I site of a plasmid vector. What information CANNOT be retrieved from this library?

A. Gene regulatory sequences.
B. Intron sequences.
C. The sequences of the telomeres (the ends of the chromosomes).
D. Amino acid sequences of proteins.

10–17   Easy, short answer

Why is a mixture of probes used to search for a gene in a DNA library when the only information on the DNA sequence of the gene has been obtained from the amino acid sequence of the protein it encodes?

cDNA Libraries Represent the mRNA Produced by a Particular Tissue (Pages 329–331)

10–18   Easy, multiple choice

You have an oligonucleotide probe that hybridizes to part of gene A. In which of the following instances would you use a cDNA library rather than a genomic library to clone gene A?

    A. You want to find both gene A and genes located near gene A on the chromosome.

    B. You want to determine the amino acid sequence of the protein encoded by gene A.

    C. You want to study the alternative splicing of gene A RNAs.

    D. Gene A may encode a ribosomal RNA.

    E. Gene A may encode a tRNA.

10–19   Easy, short answer

What is the main reason for using a cDNA library rather than a genomic library to isolate a human gene from which you wish to make large quantities of the human protein in bacteria?

10–20   Difficult, matching/fill in blanks

Some clones from cDNA libraries can have defects because of the way a cDNA library is constructed. For each of the cases 1–4, indicate which of the problems a–d you might encounter.

    1. The mRNA corresponding to the clone you are looking for was degraded at its 5′ end by a nuclease.

    2. The mRNA corresponding to the clone you are looking for was degraded at its 3′ end by a nuclease.

    3. The 5′ end of the reverse transcriptase product of the gene you are trying to clone hybridizes to sequences in the middle of the gene.

    4. The gene you are trying to clone has a long stretch of A's in the middle of the coding sequence.

    a. The 5′ part of the gene will be missing.

    b. The 3′ part of the gene will be missing.

    c. An internal fragment of the gene will be missing.

    d. The gene will be missing from the library.

Hybridization Allows Even Distantly Related Genes to Be Identified (Pages 331–332)

10–21   Intermediate, short answer

You have cloned the cDNA corresponding to a human gene encoding a membrane protein involved in cell-cell signaling and want to see if this protein is found in other organisms. A friend down the hall gives you one of his "zoo blots," which he has made by isolating DNA from various organisms, cleaving the DNA with a restriction nuclease, running the DNAs out on an agarose gel, and blotting the DNA onto a piece of filter paper. You probe the zoo blot with the 20-bp oligonucleotide that you used to clone the gene from your human library and get the result shown in Figure Q10–21.

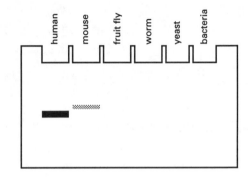

From this result, it looks as though most of these organisms may not have a gene related to your human gene. What two changes could you make to your procedure to increase your chances of finding related genes, if they exist, in the other organisms?

Q10–21

The Polymerase Chain Reaction Amplifies Selected DNA Sequences (Pages 332–335)

10–22   Intermediate, multiple choice (Requires information from Chapter 6)

If you used the radioactively labeled DNA primer shown in Figure Q10–22 to prime *in vitro* DNA synthesis by DNA polymerase on the template shown, in what form would the radioactive label (indicated by the asterisk) be at the end of your experiment?

A. Incorporated into the double-stranded DNA product.

B. As primer (i.e., nothing happens).

C. dCTP.

D. dCMP.

E. Phosphate.

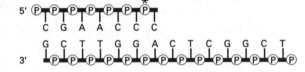

Q10–22

10–23   Intermediate, multiple choice + short answer

PCR uses a special heat-stable DNA polymerase (Taq polymerase) that is slightly less accurate than DNA polymerase (Pol I) purified from *E. coli*. Taq polymerase will therefore introduce the wrong base into a growing DNA chain more frequently than will Pol I. In which one or more of the following applications of PCR will this type of inaccuracy be a problem? Explain your answer.

A. DNA fingerprinting.

B. Detection of viral RNA.

C. Amplification of a gene from genomic DNA in order to clone it into a bacterial vector.

D. Amplification of a gene from cDNA in order to clone it into a bacterial vector.

10–24   Easy, short answer

Why is a heat-stable DNA polymerase from a thermophilic bacterium (the Taq polymerase) used in the polymerase chain reaction rather than a DNA polymerase from *E. coli* or humans?

10–25   Easy, multiple choice

Which of the following is a limitation on the use of PCR to detect and isolate genes?

    A. The sequence at the beginning and end of the DNA to be amplified must be known.

    B. It also produces large numbers of copies of sequences beyond the 5′ or 3′ end of the desired sequence.

    C. It cannot be used to amplify a particular sequence from a mixture of mRNAs.

    D. It cannot be used to amplify cDNAs.

    E. It will amplify only sequences present in multiple copies in the DNA sample.

10–26   Difficult, multiple choice

Your friend who works in a medical laboratory is distraught because he is certain that he has contracted a virus from handling patients' blood samples. He shows you a photograph of a gel on which he has run PCR products made using his own blood, blood from three patients suffering from the viral disease, or a leaf from his petunia (Figure Q10–26). The primers he added to his PCR reaction represented sequences in the viral genome.

You advise your friend not to panic and suggest that he repeat his experiment before making any rash, life-altering decisions. In addition, which of the following modifications in PCR procedure would you suggest that your friend make?

    A. Use a new tube of polymerase.

    B. Use a higher temperature while hybridizing the primers.

    C. Increase the number of cycles.

    D. Increase the concentration of primers.

    E. Decrease the amount of time in each cycle during which the polymerase is allowed to synthesize DNA.

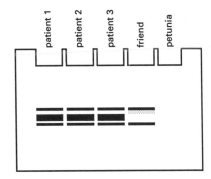

Q10–26

DNA ENGINEERING (Pages 335–342)
Completely Novel DNA Molecules Can Be Constructed (Pages 335–337)

10–27   Intermediate, multiple choice

Using a combination of magnets and iron beads you have purified a bacterial protein that binds tightly to iron. You are quite proud of this feat, but your advisor reminds you that you still haven't purified the *Drosophila* protein whose gene you cloned last year. The purification of your iron-binding protein was so simple that you hit upon the brilliant idea of fusing the coding sequences for the *Drosophila* protein to the coding sequences for your bacterial iron-binding protein and purifying the resulting fusion protein from bacteria using your magnetic scheme. Which of the following problems might you encounter?

    A. Most bacterial vectors cannot tolerate more than one DNA insert.

    B. *Drosophila* DNA cannot be ligated to bacterial DNA.

    C. Genes cloned from *Drosophila* will mutate rapidly in *E. coli.*

    D. Hybrid DNA molecules cannot be transcribed.

    E. The fusion protein may not bind to iron any longer.

Rare Cellular Proteins Can Be Made in Large Amounts Using Cloned DNA (Pages 337–338)

10–28   Easy, short answer

Insulin is a small protein hormone that regulates blood sugar level, and is given by injection to people who suffer from the disease diabetes. Diabetics used to use insulin purified from pig pancreas to control their diabetes. Give two reasons why the drug companies who produce insulin wanted to clone the human insulin gene when they already had a reasonably cheap source of the hormone.

RNAs Can Be Produced by Transcription *in Vitro* (Pages 338–339)

10–29   Easy, multiple choice

Scientists produce large quantities of RNA by transcription *in vitro* rather than by expression of cloned genes *in vivo* primarily because:

    A. *in vitro* transcription removes the need to clone the gene for the RNA into a DNA vector.

    B. the viral RNA polymerases used for *in vitro* transcription are not active *in vivo.*

    C. RNA molecules are present in only a few copies per cell and thus only small amounts of RNA can be prepared *in vivo.*

    D. *in vitro* transcription eliminates the need to purify the RNA produced away from the other RNA molecules in a cell.

Mutant Organisms Best Reveal the Function of a Gene (Pages 339–340)
Transgenic Animals Carry Engineered Genes (Pages 340–342)

10–30   Intermediate, multiple choice

You have been hired to create a cat that will not cause allergic reactions in cat-lovers. Your co-workers have cloned the gene encoding a protein found in cat saliva, expressed the protein in bacteria, and shown that it causes violent allergic reactions in people. But you soon realize that even if you succeed in making a knockout cat lacking this gene, anyone who buys one will easily be able to make more hypoallergenic cats just by breeding them. Which of the following will insure that people will always have to buy their hypoallergenic cats from you?

A. Injecting the modified ES cells into embryos that have a genetic defect that will prevent the mature adult from reproducing.

B. Implanting the injected embryos into a female cat that is sterile due to a genetic defect.

C. Selling only the offspring from the first litter of the female cat implanted with the injected embryos.

D. Surgically removing the sexual organs of all the knockouts before you sell them.

E. Selling only male knockouts.

Answers

A10–1. D.

A10–2. A. A nuclease hydrolyzes the <u>phosphodiester</u> bonds in a nucleic acid.

B. Nucleases that cut DNA only at specific short sequences are known as <u>restriction nucleases</u>.

C. DNA composed of sequences from different sources is known as <u>recombinant DNA</u>.

D. <u>Genetic engineering</u> can create organisms with combinations of genes that probably have never occurred naturally.

E. Millions of copies of a DNA sequence can be made entirely *in vitro* by <u>the polymerase chain reaction.</u>

A10–3. B.

A10–4. D. A restriction enzyme that has a 4 base-pair recognition sequence cuts on average once every 4^4 or 256 base pairs; one that has a 6 base-pair recognition sequence cuts once every 4^6 or 4096 base pairs; one that has an 8 base-pair recognition sequence cuts once every 4^8 or 65,536 base pairs; one that has a 12 base-pair recognition sequence cuts once every 4^{12} or 16 million base pairs. So to obtain fragments of around 70 kb in size, you would cut with an enzyme that recognizes an 8 base-pair site.

A10–5. (A) You would first digest your sample with a combination of restriction enzymes selected so that they give a set of fragment sizes that could have come from only one of the plasmids. Then you would run the resulting mixture of DNA fragments on a gel alongside a set of size markers and determine the sizes of each fragment. By looking at the restriction maps, you should then be able to match your results to one of the plasmids. (B) Digestion with any of the following combinations will enable you to distinguish which plasmid you have: Hind III + Bgl II; Eco RI + Bgl II; Eco RI + Bgl II + Hind III. The plasmids are the same size, so you cannot distinguish them simply by making a single cut (with Hind II) and determining the size of the complete DNA by gel electrophoresis. Nor can you distinguish them by cutting with all four restriction nucleases, since the set of fragment sizes produced from both plasmids will be the same. Cutting with Bam HI or Eco RI on their own is not sufficient because you will get bands of the same size from both plasmid A and plasmid B. The only difference between the two plasmids is the location of the Bgl II site relative to the two Bam HI sites, so if you cut with an enzyme that cuts outside the Bam HI fragment and with Bgl II, you will get different-sized fragments from the two plasmids.

A10–6. The sequence is 5′-ATGTCAGTCCAG-3′. You start at the bottom of the gel and read off the bands in strict order of increasing size, working across all lanes.

A10–7. D. If some of the DNA templates you are sequencing are cut at one specific site (as would be the case if it were cut by a restriction enzyme), the polymerase will stop when it comes to the end of the DNA, giving rise to at least some product of one particular size in all the reaction mixtures. So all four lanes will have a band of this particular size. In addition, you would get normal sequence from the full-length templates, and normal sequence from those templates in which the polymerase incorporated a dideoxynucleotide before encountering the end. The other options are incorrect: if you added all four dideoxynucleotides to one of the reactions (A), that lane would have a band at every position because the polymerase would stop at A's, C's, G's, and T's instead of at only one type of nucleotide. If you forgot to add deoxynucleotides to the reactions (B), you would not get any polymerization, and all of your lanes would be blank. If you forgot to add dideoxynucleotides (C), the reactions would not stop until the end of the DNA fragment, and all of the products would be full-length and would all be at the top of the gel. If your primer hybridized to more than one part of the fragment of DNA you were sequencing (E), your gel would look as though two different sequences had been superimposed onto each other.

A10–8.   A. The technique of _in situ_ hybridization can be used to detect a specific DNA sequence on a chromosome.

B. Northern blotting detects a specific sequence in <u>RNA</u>.

C. Southern blotting detects a specific sequence in <u>DNA</u>.

D. A short single-stranded DNA is an <u>oligonucleotide</u>.

E. A piece of DNA used to detect a specific sequence in a nucleic acid by hybridization is known as a <u>probe</u>.

A10–9.   (A) All three are affected. (B) The two parents have a single defective copy of the gene; the fetus has two defective copies. (C) See Figure A10–9. The two blots examined together show that Caril Ann and Charles each have one full-length copy of gene X (the bands at the top of the gel), which hybridizes with both probe A and probe B. The fetus does not. The blot with probe A shows that Caril Ann and the fetus have a short fragment of gene X that hybridizes with probe A only, indicating that this copy of gene X has a deletion somewhere other than in the region recognized by probe A. The second blot (with probe B) shows that Charles and the fetus have a short fragment that hybridizes with probe B, indicating that this copy of gene X has a deletion somewhere other than in the region recognized by probe B. Since the shortened gene found in Charles does not show up on the probe A blot, this deletion must be in the region of A; similarly, since the shortened gene found in Caril Ann does not show up on the probe B blot, her deletion must be in the region of B. The fetus has inherited two defective copies of gene X, one from each parent.

Charles has one full-sized copy of gene X and one copy with a deletion in region A

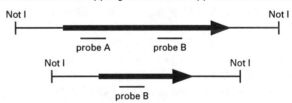

Ann has one full-sized copy of gene X and one copy with a deletion in region B

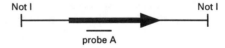

The fetus has one copy of gene X with a deletion in region A and one copy with a deletion in region B

A10–9

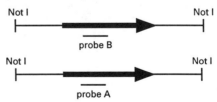

A10–10. 1, all four; 2, all four; 3, B and D. Both the upper and lower strands of DNA are present in genomic Southern blots, so all four oligos will hybridize to either Southern blot. (Oligonucleotides A and B will still be able to hybridize to genomic DNA cut with Bgl II, since they can still base-pair to the individual fragments that result from the digest.) Northern blots contain only RNA, which has the sequence of the upper strand of the DNA. Hence, only B and D will hybridize to a Northern blot.

A10–11. E. *In situ* hybridization can be used to determine the distribution of an mRNA in different tissues.

A10–12. A. Two fragments of DNA can be joined together by <u>DNA ligase</u>.

B. Restriction enzymes that cut DNA straight across the double helix produce fragments of DNA with <u>blunt ends</u>.

C. A fragment of DNA is inserted into a <u>vector</u> in order to be cloned in bacteria.

D. A <u>cDNA</u> library contains a collection of DNA clones derived from mRNAs.

E. A <u>genomic</u> library contains a collection of DNA clones derived from chromosomal DNA.

A10–13. (A) 1 and 2. (B) 1, 2, and 4. Blunt ends, such as those left by Sma I, can always be joined by DNA ligase only. In addition, the staggered ends of DNA cut by Sal I and Xho I fit together by base-pairing and can also be joined by ligase alone.

A10–14. A bacterial replication origin to allow the plasmid to be replicated, at least one unique restriction site to allow easy insertion of foreign DNA, and an antibiotic-resistance gene (or some other selectable marker gene) to allow selection for bacteria that have taken up the recombinant plasmids.

A10–15. C. The size of the fragments that are left after a restriction digest does not depend on the total size of the genome; it depends on the sequence of the genome and the frequency with which the restriction enzyme recognition site is found in the genome. A and B are true: as a limiting case, you can think of what would happen if a fragment the size of the entire genome inserted into the bacterial vector. In this case, you would have to screen only one colony to find the clone that hybridizes to your probe, but it will be very difficult to find out where on the insert your gene of interest lies. D is true: the larger the gene you are seeking, the more likely it is that there will be a restriction fragment in the gene (or that the gene will be broken if the DNA was fragmented by random shearing), and hence the less likely it is that the entire gene will be found in one clone. E is true: as the size of your oligo probe decreases, the chance of finding that sequence randomly in the genome increases (just as the number of restriction sites increases when the size of the recognition site decreases).

A10–16. C. The very ends of all of the chromosomes are unlikely to be Not I sites, meaning that the fragments containing the ends of the chromosomes will not be able to insert into the bacterial vector (since they have not been cut by Not I at both ends) and will be lost from the library. All sequences present in genomic DNA (which includes regulatory sequences and introns) should be present in a genomic library. The coding sequence of the gene (and hence the amino acid sequence of the encoded protein) is also present in a genomic clone, although it is interrupted by intron sequences and therefore somewhat difficult (but not impossible) to determine.

A10–17. Because most amino acids can be encoded by more than one codon, a given sequence of amino acids could be encoded by a number of different nucleotide sequences. Probes corresponding to all these possible sequences have to be synthesized in order to be sure of including the one that corresponds to the actual nucleotide sequence of the gene and thus will hybridize with it.

A10–18.   B. Although you can determine the amino acid sequence from a genomic clone, it is much easier to do so if you do not have to sequence the introns, as is the case for a cDNA clone. A is incorrect: since cDNA libraries are derived from mRNA, each clone contains only the coding sequence of the gene and a small amount of surrounding untranslated sequence. C is incorrect: you could use a cDNA clone to make RNA *in vitro*, but the RNA made would have no introns and thus could not be used to study alternative splicing. D and E are incorrect: since ribosomal RNAs and transfer RNAs do not have poly(A) tails, they are not represented in a cDNA library.

A10–19.   The gene isolated from a genomic library would still contain introns, and bacteria do not contain the biochemical machinery for removing introns by RNA splicing. The same gene isolated from the cDNA library will have already had its introns removed.

A10–20.   1 = a; 2 = d; 3 = a; 4 = b. (1) If the mRNA is degraded from the 5′ end, it would still be reverse transcribed and would end up in the library as a clone lacking its 5′ end. (2) If the mRNA was degraded from the 3′ end, it would be missing its 3′ poly(A) tail. In the construction of a cDNA library, only molecules that still have their poly(A) tail will be reverse transcribed, so mRNAs missing their 3′ end will not be represented in the library. (3) If the 5′ end hybridizes to sequences in the middle of the gene, the "hairpin" formed when the single-stranded DNA loops back on itself to form the primer for DNA polymerase will be very large. After this loop is digested, the remaining double-stranded DNA fragment will be missing the 5′ end of the gene. (4) If the gene has a long stretch of internal A's, the poly(T) primer used in the reverse transcription step could hybridize to the internal poly(A) stretch rather than to the poly(A) tail, and the resulting cDNA will have lost its 3′ end.

A10–21.   You could (1) lower the hybridization temperature and (2) use the entire human clone rather than just the oligonucleotide as your probe. The stringency of hybridization depends on the temperature; the lower the temperature, more mismatches will be tolerated in the hybridization. If two genes are related, they may be very similar in sequence in regions of the gene not covered by your original oligonucleotide probe; using the entire cDNA clone as your probe will increase the number of potential base pairs between your probe and a related gene on the zoo blot.

A10–22.   D. The radioactively labeled C is a mismatch, so the editing function of DNA polymerase will cleave off the mismatched nucleotide 5′ to the phosphate, generating radioactive dCMP and a free 3′-OH group, and allowing the rest of the top strand to be synthesized.

A10–23.   C and D. In DNA fingerprinting and the detection of viral DNA, the DNA fragments generated by PCR are used solely as a device to detect a DNA sequence, and the sequence of the PCR products per se does not matter. You are simply using the size and/or the presence or absence of a PCR product as the assay. When you use PCR to clone a gene, however, the actual DNA that is being inserted into the vector is DNA that was synthesized by the Taq polymerase. Therefore, if Taq is prone to making errors, it is likely that some of the fragments that you clone will have errors, which may be a problem if you want to know the sequence of the gene or use your PCR-derived clone to produce protein.

A10–24.   The PCR technique involves heating the reaction at the beginning of each cycle to separate the newly synthesized DNAs into single strands so that they can act as templates for the next round of DNA synthesis. Using a heat-stable polymerase avoids having to add it afresh for each round of DNA replication.

A10–25.   A. In order to construct primers that will bracket the desired gene, you have to know the sequence at the beginning and end of the DNA to be copied. Although a few copies of sequences beyond the ends of the desired sequence (B) are made in the early cycles, these soon become a negligible fraction of the total DNA synthesized. You can use PCR to amplify a particular RNA sequence (C) by using the appropriate primers to first guide the reverse transcriptase reaction that makes a DNA copy of the RNA and then to guide the synthesis of the complementary DNA strand. PCR can be used to amplify a sequence from any DNA, including cDNAs (D). PCR is extremely sensitive and can detect and amplify a particular sequence even if it is present only in a single copy in the DNA sample (E).

A10–26.   B. The absence of PCR fragments in the petunia lane tells you that there is no viral contaminant in any of your friend's general reagents, so he can use the same tube of polymerase when he repeats the experiment. One of the problems with PCR is that it is sometimes too sensitive; if the primers mis-hybridize, they will form a product that is then readily amplified in all the subsequent cycles. Thus it is possible that the thick band in the patient's samples are PCR products amplified from the viral genome while the thin (less intense) bands represent a mis-hybridization of the primers to human DNA. One way to prevent this mis-hybridization is to increase the temperature at which you hybridize the primers to the DNA. Increasing the concentration of primers will exacerbate the problem of nonspecific hybridization. Increasing the number of cycles will simply increase the amount of PCR products, both specific and nonspecific. Decreasing the amount of time that the polymerase is allowed to synthesize DNA will have no effect on primer mis-hybridization.

A10–27.   E. Because DNA is pretty much the same (except for sequence) no matter what the source, bacterial DNA can be ligated to *Drosophila* DNA, and once that happens, neither organism could tell the difference between the two parts of the chimeric molecule. Proteins, on the other hand, function according to their folding, and fold according to their sequence, so joining two sequences together into a single chain may affect the way each functions.

A10–28.   Any two of the following would be acceptable:

(1) Cloning the gene allows human insulin to be produced in large quantities from bacteria or other cells carrying the cloned DNA sequence.

(2) It is easier and less costly to extract the same amount of insulin from a bacterial culture than from pig pancreas.

(3) Insulin made in a bacterial culture and then purified will be free of any possible contaminating viruses that pigs (and any other whole animal) tend to harbor.

(4) The pig protein has slight differences from the human protein, which can lead to side effects on prolonged use. Wherever possible, a human protein would be preferred for clinical treatment of this sort.

A10–29.   D. In order to produce an RNA by transcription *in vitro*, you must first clone the DNA of the gene you wish to transcribe in order to get a large amount of pure template. Many RNAs are produced at high levels in cells. Viral RNA polymerases are able to transcribe RNA *in vivo* to especially high levels, since their purpose is to make high levels of viral proteins.

A10–30.   D. If you do (A), you will not be able to make any knockout cats because the first litter (which will at best have a few mosaics in which one copy of the gene has been knocked out in the germ cells) will be sterile and you will not be able to mate them. The genotype of the female cat in which you implant the embryos has no effect on the genotype of the embryos (B). Since the first litter is mosaics, they will not be hypoallergenic, since they will still have one or two copies of the gene in their cells (C). Selling only male knockouts does no good, because they can be mated to normal females and the heterozygous offspring can be backcrossed (D). Neutering all the knockout animals that you sell is the only option of the five listed that will prevent happy pet owners from becoming happy pet breeders.

11 Membrane Structure

Questions

THE LIPID BILAYER (Pages 348–357)

11–1   Easy, matching/fill in blanks

For each of the following sentences, fill in the blanks with the correct word or phrase selected from the list below. Use each word or phrase only once.

   A. The specialized functions of different membranes are largely determined by the _____ they contain.

   B. Membrane lipids are _____ molecules, composed of a hydrophilic portion and a hydrophobic portion.

   C. All cell membranes have the same _____ structure, with the _____ of the phospholipids facing into the interior of the membrane and the _____ on the outside.

   D. The most common lipids in most cell membranes are the _____.

   E. The head group of a glycolipid is composed of _____.

lipid bilayer; lipid monolayer; proteins; lipids; hydrophobic; phospholipids; amphipathic; amphoteric; sterols; glycolipids; fatty acid tails; hydrophilic head groups; phosphatidylcholine; cholesterol; phosphatidylserine; sugars.

Membrane Lipids Form Bilayers in Water (Pages 349–352)

11–2   Easy, multiple choice

Which of the following membrane lipids do not contain a fatty acid tail?

   A. Phosphatidylcholine.
   B. A glycolipid.
   C. Phosphatidylserine.
   D. Sphingomyelin.
   E. Cholesterol.

11–3   Easy, multiple choice

The lipid bilayer is held together mainly by:

   A. covalent bonding between membrane lipids.
   B. hydrogen-bonding between the phospholipid tails.
   C. repulsion between the phospholipid tails and water.
   D. covalent bonding between the ends of phospholipid tails in opposite layers.
   E. hydrogen-bonding between the head groups of phospholipids.

The Lipid Bilayer Is a Two-dimensional Fluid (Pages 352–353)

11–4   Intermediate, multiple choice (Requires information from sections on pages 349–352 and pages 360–361)

Which of the following statements regarding lipid membranes are true?

    A. Phospholipids will spontaneously form liposomes in nonpolar solvents.

    B. A solution of pure fatty acids forms a lipid bilayer in a polar solvent.

    C. Membrane lipids move laterally within their own layer.

    D. Membrane lipids frequently move between one layer of the bilayer and the other.

    E. The preferred form of a lipid bilayer in water is a flat sheet with exposed edges.

The Fluidity of a Lipid Bilayer Depends on Its Composition (Pages 353–354)

11–5   Intermediate, multiple choice

A bacterium is suddenly expelled from a warm human intestine into the cold world outside. Which of the following adjustments might the bacterium make to maintain the same level of membrane fluidity?

    A. Increase the length of the hydrocarbon tails in its membrane phospholipids.

    B. Increase the proportion of unsaturated hydrocarbon tails in its membrane phospholipids.

    C. Increase the proportion of hydrocarbon tails with no double bonds in its membrane phospholipids.

    D. Decrease the amount of cholesterol in the membrane.

    E. Decrease the amount of glycolipids in the membrane.

11–6   Intermediate, multiple choice (Requires information from Chapter 2)

Which of the following statements regarding the fatty acid tails of phospholipids are false?

    A. Phospholipids with unsaturated tails make the bilayer more fluid because the tails contain fewer hydrogens and thus form fewer hydrogen bonds with each other.

    B. Saturated phospholipid tails pack more tightly against each other than do unsaturated tails.

    C. Most membrane phospholipids have one fully saturated tail.

    D. Phospholipid tails in a membrane can interact with each other via van der Waals interactions.

    E. Fatty acid tails vary in length.

The Lipid Bilayer Is Asymmetrical (Pages 354–355)

11–7   Easy, multiple choice

New membrane synthesis occurs by:

    A. the spontaneous aggregation of free phospholipids into a new bilayer in the aqueous environment of the cell.

    B. incorporation of phospholipids into both faces of a preexisting membrane by enzymes attached to each face.

    C. incorporation of phospholipids into one face of a preexisting membrane and their random redistribution to both faces by flippases.

    D. incorporation of phospholipids into one face of a preexisting membrane and their specific redistribution by flippases.

11–8   Intermediate, multiple choice (Requires information from section on page 358)

The plasma membrane of an animal cell is symmetric with regard to:

    A. the distribution of different phospholipids in each half of the lipid bilayer.

    B. the distribution of cholesterol in each half of the lipid bilayer.

    C. the orientation of membrane proteins in the lipid bilayer.

    D. the distribution of glycolipids in the lipid bilayer.

    E. All of the above.

11–9   Intermediate, multiple choice

Three phospholipids X, Y, and Z are distributed in the plasma membrane as indicated in Figure Q11–9. For which of these phospholipids does a flippase probably exist?

    A. X only.

    B. Z only.

    C. X and Y.

    D. Y and Z.

    E. X and Z.

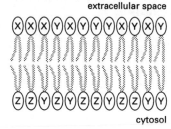

Q11–9

Lipid Asymmetry Is Generated Inside the Cell (Pages 355–356)

11–10   Easy, multiple choice

Where does most new membrane synthesis take place in a eucaryotic cell?

    A. The Golgi apparatus.

    B. The endoplasmic reticulum.

    C. The plasma membrane.

D. The mitochondria.

E. Cytoplasmic membrane vesicles.

11-11   Intermediate, short answer (Requires general knowledge of protein terminology and information from sections on pages 357–367)

A small membrane vesicle in the cytoplasm of a eucaryotic cell fuses with the plasma membrane.

(A) Sketch the membrane vesicle before fusion, indicating the cytosolic and non-cytosolic faces of its membrane.

(B) Sketch the plasma membrane after vesicle fusion, indicating the new location of the vesicle membrane and labeling its original cytosolic and noncytosolic faces. Label the extracellular space and the cytoplasm.

(C) If the vesicle carried a transmembrane protein with its amino terminus facing into the interior of the vesicle and its carboxyl terminus facing the cytosol, indicate on your sketch in B the orientation of this protein after vesicle fusion with the plasma membrane.

11-12   Easy, multiple choice

Why are glycolipids found on the extracellular but not the cytoplasmic surface of the plasma membrane?

A. Flippases transport them from the cytosolic face.

B. The enzymes that produce them are present only on the extracellular surface of the plasma membrane.

C. The enzymes that produce them are present only on the noncytosolic side of the Golgi apparatus, and there are no flippases that can flip glycolipids to the cytosolic face of the lipid bilayer.

D. The oligosaccharides on glycolipids are cleaved off by enzymes found only in the cytosol.

E. They flip spontaneously after incoroporation due to the hydrophilic sugar head groups.

Lipid Bilayers Are Impermeable to Solutes and Ions (Pages 356–357)

11-13   Intermediate, multiple choice

A hungry yeast cell lands in a vat of grape juice and begins to feast on the sugars there, producing carbon dioxide and ethanol in the process:

$$C_6H_{12}O_6 + 2ADP + 2P_i + H^+ \rightarrow 2CO_2 + 2CH_3CH_2OH + 2ATP + 2H_2O$$

Unfortunately, the grape juice is contaminated with proteases, enzymes that attack some of the transport proteins in the yeast cell membrane, and the yeast cell dies. Which of the following could account for the yeast cell's demise?

A. Toxic buildup of carbon dioxide inside the cell.

B. Toxic buildup of ethanol inside the cell.

C. Diffusion of ATP out of the cell.

D. Inability to import sugar into the cell.

E. Inability to take water into the cell.

MEMBRANE PROTEINS (Pages 357–367)
Membrane Proteins Associate with the Lipid Bilayer in Various Ways (Page 358)

11–14   Easy, multiple choice

How do cells change the orientation of proteins in the plasma membrane?

A. By increasing the fluidity of the lipid bilayer.

B. By using enzymes which rotate proteins through the bilayer.

C. By budding and fusing of membrane vesicles.

D. By temporary breaks in the bilayer that release the proteins from the membrane.

E. The orientation of membrane proteins cannot be changed.

A Polypeptide Chain Usually Crosses the Bilayer as an α Helix (Pages 358–360)

11–15   Intermediate/difficult, multiple choice (Requires information from Chapter 5)

A group of membrane proteins, which can be extracted only from membranes using detergents, are all found to have a similar amino acid sequence at their carboxyl terminus: -KKKKXXC (where K stands for lysine, X stands for any amino acid and C stands for cysteine). This sequence is essential for their attachment to the membrane. What is the most likely way in which the carboxyl-terminal sequence attaches these proteins to the membrane?

A. The cysteine residue is covalently attached to a membrane lipid.

B. The peptide spans the membrane as an α helix.

C. The peptide spans the membrane as part of a β sheet.

D. The positively charged lysine residues interact with an acidic integral membrane protein.

11–16   Intermediate, multiple choice

A strain of bacteria secretes a toxin that can lyse human red blood cells. You are able to partially purify the toxin and find that it is a small protein. Furthermore, the toxin is capable of rendering liposomes made of pure phospholipids permeable to many different ions. What type of protein is the bacterial toxin likely to be?

A. A flippase.

B. A β-barrel protein.

C. A protease.

D. A protein containing a single hydrophobic α helix.

E. An enzyme that adds carbohydrate groups to lipids.

Membrane Proteins Can Be Solubilized in Detergents and Purified (Pages 360–361)

11–17   Intermediate, multiple choice (Requires information from section on pages 358–360 and information from Chapter 5)

Which of the following statements regarding membrane proteins are false?

    A. Integral membrane proteins often precipitate (form insoluble aggregates) in aqueous solutions lacking detergents.

    B. Some hydrophobic amino acids in membrane proteins are not in contact with the lipid bilayer.

    C. A membrane-spanning α helix composed of hydrophobic amino acids will often unfold in water.

    D. Strong detergents can completely unfold both membrane and nonmembrane proteins.

    E. In transmembrane proteins that form an aqueous pore through the membrane, the pore is lined with hydrophobic amino acid side chains.

The Complete Structure Is Known for Very Few Membrane Proteins (Pages 361–363)

11–18   Intermediate/difficult, short answer (Requires information from Chapter 2)

In order to investigate the action of a bacterial membrane protein that is a light-driven proton pump, you purify the protein and assemble it together with phospholipids into liposomes that also contain an indicator dye, which is blue at high pH, colorless at neutral pH, and red at low pH. You then expose the liposomes, in an aqueous solution, to sunlight. The interior of the liposomes stays colorless. Assuming that your bacterial protein has not been damaged by its purification and incorporation into liposomes, what is the most likely explanation for this result?

11–19   Intermediate/difficult, data interpretation

On further investigation of the proton pump protein described in Question 11–18, you find that treatment of either intact bacteria or the purified protein with the protease chymotrypsin results in cleavage of the protein at a single site near the amino terminus, which destroys its capacity to pump protons. When you treat the liposomes described in Question 11–18 with chymotrypsin, the interior of the liposomes turns red. From these results, which of the diagrams in Figure Q11–19 (*next page*) illustrates the orientation of the pump protein in the plasma membrane of the intact bacterium and the direction in which it pumps protons? Explain your reasoning.

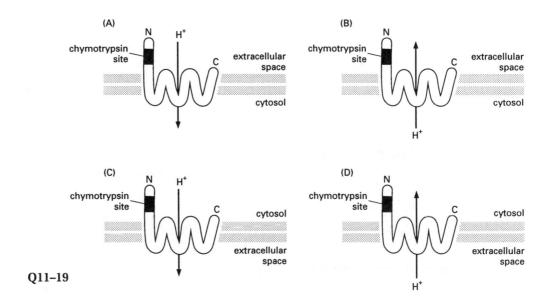

Q11–19

The Plasma Membrane Is Reinforced by the Cell Cortex (Pages 363–364)

11–20   Easy, multiple choice

Which of the following functions does the cell cortex perform?

    A. It influences the passage of small molecules into and out of the cell.

    B. It enables cells to change shape.

    C. It lubricates the cell.

    D. It restricts the movement of certain proteins in the lipid bilayer.

    E. It supports and strengthens the membrane.

11–21   Intermediate/difficult, short answer (Requires information from sections on pages 348–357 and information from Chapter 5)

You have isolated two mutants of a normally pear-shaped microorganism that have lost this distinctive shape and are now round. One of the mutants has a defect in a protein you call A and the other a defect in a protein you call B. You grind up each type of mutant cell and normal cells separately and separate the plasma membranes from the cytoplasm. You then wash the membrane fraction with urea and centrifuge the mixture; the membranes form a pellet, while the proteins liberated from the membranes by the wash remain in the supernatant. When you check each of the fractions for the presence of A or B, you obtain the results given below.

	First cell extract		After urea wash and centrifugation	
	Membrane	Cytosol	Membrane	Supernatant
Normal cells	A and B	no A or B	A	B
Mutant A	A	B	A	no A or B
Mutant B	A	B	A	no A or B

Which of the following statements are consistent with your results?

    A. Protein A is an integral membrane protein that interacts with B, a peripheral membrane protein that is part of the cell cortex.

    B. Protein B is an integral membrane protein that interacts with A, a peripheral membrane protein that is part of the cell cortex.

    C. Protein A is an integral membrane protein that interacts with B, a peripheral membrane protein located on the exterior of the cell.

    D. The mutation in A affects its ability to interact with B.

    E. The mutation in A affects its ability to interact with the membrane.

The Cell Surface Is Coated with Carbohydrate (Pages 364–366)

11–22   Intermediate, multiple choice

Sperm cells bind to an egg via ZP3, a glycoprotein in the glycocalyx of the egg, and can be inhibited from doing so by adding an excess of free oligosaccharide. You are interested in finding the protein on the sperm cell responsible for binding to ZP3. Which of the following is it most likely to resemble?

    A. Spectrin.

    B. A lectin.

    C. A porin.

    D. Bacteriorhodopsin.

    E. A proteoglycan.

11–23   Intermediate, multiple choice

Diversity among the oligosaccharide chains found in the glycocalyx can be achieved in which of the following ways?

    A. Varying the types of sugar monomers used.

    B. Varying the types of linkages between sugars.

    C. Varying the number of monomers in the chain.

    D. Varying the number of branches in the chain.

    E. All of the above.

Cells Can Restrict the Movement of Membrane Proteins (Pages 366–367)

11–24   Easy, data interpretation

Four samples of cells have membranes containing a uniform distribution of a fluorescently labeled phospholipid (sample 1) and three fluorescently labeled membrane proteins X, Y, and Z (samples 2–4), respectively. An intense beam of light is shone on each sample. The light beam destroys the fluorescent label in the illuminated area, but the intensity of fluorescence in the area recovers over time, as bleached molecules diffuse away and unbleached molecules diffuse in. The results of this experiment are shown in Figure Q11–24. Which of the following interpretations are INCONSISTENT with these results?

    A. X is a small freely diffusing protein.

B. Y is a large freely diffusing protein.

C. X and Y are part of the same large protein complex.

D. Z is bound to a component of the cell cortex.

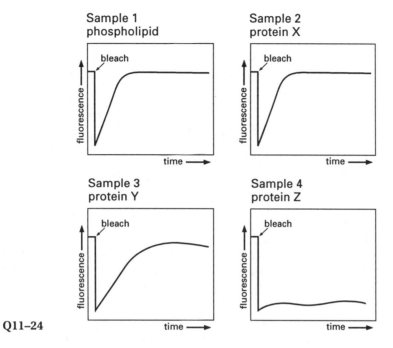

Q11–24

11–25   Intermediate, multiple choice (Requires student to have studied the whole chapter)

Which of the following movements can a membrane protein make by lateral diffusion through the lipid bilayer?

A. Movement from the apical plasma membrane to the basal plasma membrane in a gut epthelial cell.

B. Movement from the outer nuclear membrane to the inner nuclear membrane.

C. Movement from the outer mitochondrial membrane to the inner mito-chondrial membrane.

D. Movement from the outer nuclear membrane to the endoplasmic reticulum.

E. None of the above.

Answers

A11–1. A. The specialized functions of different membranes are largely determined by the <u>proteins</u> they contain.

B. Membrane lipids are <u>amphipathic</u> molecules, composed of a hydrophilic portion and a hydrophobic portion.

C. All cell membranes have the same <u>lipid bilayer</u> structure, with the <u>fatty acid tails</u> of the phospholipids facing into the interior of the membrane and the <u>hydrophilic head groups</u> on the outside.

D. The most common lipids in most cell membranes are the <u>phospholipids</u>.

E. The head group of a glycolipid is composed of <u>sugars</u>.

A11–2. E.

A11–3. C.

A11–4. C. The remaining answers are false. Phospholipids form bilayers only in polar solvents (A). Fatty acids, because of their single tail, form micelles not bilayered structures in water (B). Membrane lipids rarely move between one layer of the bilayer and the other (D). The preferred form of a lipid bilayer in water is a closed sphere to avoid contact of the hydrophobic groups at the edges of the bilayer with water (E).

A11–5. B. At colder temperatures, the membrane will be less fluid. Hence, in order to maintain the status quo, the bacterium will have to take measures to increase membrane fluidity. Increasing the length of the hydrocarbon tails (A) would decrease membrane fluidity, while decreasing the number of glycolipids (E) would have little or no effect. Decreasing the proportion of fatty acid tails with no double bonds (fully saturated) (C) would decrease membrane fluidity. Bacteria do not have cholesterol (D).

A11–6. A. Unsaturated fatty acid tails do have fewer hydrogens and do interact less well with one another but not for the reason stated. The decrease in interaction is due to a decrease in van der Waals interactions between the hydrocarbon tails because they can pack less closely. Hydrocarbon chains cannot form hydrogen bonds with each other.

A11–7. D.

A11–8. B. All other membrane components are distributed asymmetrically, being more common in, or being confined to, one layer of the bilayer. Membrane proteins each have a unique orientation in the membrane.

A11–9. C. As phospholipids are initially inserted into the cytosolic face of the lipid bilayer, flippases would be required to move X and Y to the opposite face, as they would not spontaneously move.

A11–10. B.

A11–11. (A) Figure A11–11A.
(B and C) Figure A11–11B.

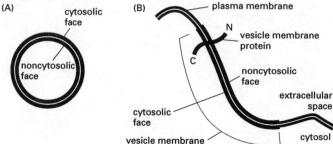

A11–11

A11–12. C.

A11–13. D. Because the lipid bilayer is permeable to carbon dioxide and ethanol, destroying membrane proteins is unlikely to affect their exit (A and B). The lipid bilayer is also permeable to water (E). On the other hand glucose requires a membrane transport protein to be imported into the cell. ATP, which is a nucleotide, also requires a transport protein to cross a membrane and thus could not be lost from the cell by simple diffusion (C).

A11–14. E.

A11–15. A. The peptide is extremely hydrophilic and is therefore unlikely to be inserted into the lipid bilayer. It is also too short to span the membrane as an α helix. While it is possible that the lysines interact with an acidic membrane protein, if such an interaction were solely responsible for attaching the protein to the membrane, it would not require a detergent to remove the protein, since ionic interactions are disrupted by milder treatments such as salt washes and pH changes.

A11–16. B. Insertion of a pore-forming protein into the lipid bilayer will have the effects noted by making the red blood cell unable to regulate its internal ion composition. The type of protein most likely to form the sort of nonspecific pore described is a β-barrel protein. Enabling membrane lipids to flip from one layer of the bilayer to the other should not affect permeability (A). A protease would have no effect on liposomes made of pure phospholipids (C). Proteins containing a single hydrophobic α helix would not form a channel through the membrane (D). Addition of carbohydrate groups to lipids should not make the bilayer any more permeable to ions (E).

A11–17. E. In transmembrane proteins that form an aqueous pore through the membrane, the pore is lined with hydrophilic amino acid side chains. The other statements are all true. Integral membrane proteins often precipitate in aqueous solutions because of their stretches of hydrophobic amino acids (A). Proteins also contain hydrophobic amino acids in parts of the protein other than the membrane-spanning region, for example, in the cores of their extracellular or cytoplasmic domains (B). A membrane-spanning α helix composed of hydrophobic amino acids will unfold in water as the hydrophilic parts of the polypeptide backbone make hydrogen bonds with the water molecules instead of with each other (C).

A11–18. The bacterial membrane protein has become incorporated into the liposomes with roughly equal numbers of proteins in opposite orientations. Thus while the proteins in one orientation are pumping protons out of the liposome interior, the proteins in the other orientation are pumping protons into the liposome interior. The liposome interior therefore stays at a neutral pH.

A11–19. B. Because chymotrypsin can cut the proton pump protein in an intact bacterium, the amino terminus of the pump protein must be on the extracellular side of the bacterial plasma membrane. Treatment of the liposomes with chymotrypsin will destroy all of the pumps oriented so that their amino termini are exposed on the outside of the liposome. Pump proteins with their amino termini on the inside of liposome will be protected. Since the liposome interiors then turn red, the pH in the interior must have decreased, meaning that protons are being pumped in. Thus the proton pump must pump protons toward the face on which the amino terminus is exposed. In an intact bacterial cell, that will be from the cytosol to the outside.

A11–20. A, D, and E.

A11–21. A and D. The results from the extracts of normal cells show that protein A is an integral protein that remains in the membrane through all the treatments, while protein B is a peripheral protein that can be removed from the membrane by urea. In the cell extracts from the mutants in

A, the protein A still remains in the membrane, but the B protein does not. This is consistent with the mutation in A affecting its interaction with B. The same results are obtained when the B protein is mutant, consistent with the idea that A and B interact. The loss of an interaction between an integral membrane protein and a protein in the cortex would be more likely to result in a change in cell shape than the loss of an interaction between an integral membrane protein and a protein on the exterior of the cell.

A11–22.   B. Since binding of ZP3 to its receptor is prevented by the addition of excess oligosaccharide, which competes with the receptor for ZP3 binding, the receptor must recognize and bind oligosaccharide strongly. The sperm receptor is thus likely to bind to ZP3 by recognizing the carbohydrate groups attached to it. Thus it is most likely that the receptor is a carbohydrate-binding protein, a lectin.

A11–23.   E.

A11–24.   C. The graphs show that X diffuses almost as quickly as a lipid, Y diffuses more slowly (as would be expected if Y were larger than X), and that Z hardly diffuses at all (as would be expected of a protein bound to the extracellular matrix or the cell cortex). If X and Y were part of the same complex, one would expect their rates of diffusion to be the same (i.e., the graphs for X and Y should look identical).

A11–25.   D. The outer nuclear membrane and the endoplasmic reticulum are continuous so that a protein can diffuse within the membrane from one compartment to the other. Movement from the apical to basal plasma membrane in gut epithelial cells is blocked by the tight junctions (A). Movement from the outer nuclear membrane to the inner nuclear membrane is blocked by the nuclear pores (B). The outer mitochondrial membrane is not continuous with the inner mitochondrial membrane (C).

12 Membrane Transport

Questions

The Ion Concentrations Inside a Cell Are Very Different from Those Outside (Pages 372–373)

12–1   Easy, multiple choice

The most abundant intracellular cation is:

   A. Na^+.
   B. Ca^{2+}.
   C. H^+.
   D. K^+.
   E. positively charged macromolecules.

CARRIER PROTEINS AND THEIR FUNCTIONS (Pages 373–385)
Solutes Cross Membranes by Passive or Active Transport (Page 375)

12–2   Easy, matching/fill in blanks

For each of the following sentences, fill in the blanks with the correct word selected from the list below. Use each word only once.

   A. Carrier proteins and channel proteins can provide a _____ pathway through the membrane for specific polar solutes or inorganic ions.
   B. A substance is transported down its concentration gradient by _____ transport.
   C. A substance is transported up its concentration gradient by _____ transport.
   D. _____ proteins can transport a substance down its concentration gradient only.
   E. _____ proteins are highly selective in the solutes they transport, binding the solute at a specific site.

channel; hydrophilic; passive; hydrophobic; active; carrier.

12–3   Easy, multiple choice (Requires information from section on pages 372–373)

In a typical animal cell, which of the following types of transport occur through a channel protein?

   A. Movement of amino acids into a cell.
   B. Movement of Na^+ out of a cell.
   C. Movement of Na^+ into a cell.
   D. Movement of glucose into a starved cell.
   E. Movement of glucose out of a starved cell.

Electrical Forces as Well as Concentration Gradients Can Drive Passive Transport (Pages 375–377)

12–4   Easy, matching/fill in blanks

For each of the following sentences, fill in the blanks with the correct word or phrase selected from the list below. Use each word or phrase only once.

A. The net force driving a charged solute across a cell membrane is the _____
    This is composed of two forces, one due to the _____ of the solute and the
    other due to the _____ across the membrane, which is known as the
    _____.

B. In most circumstances the inside of the cell is _____ charged with respect to
    the outside.

C. The membrane potential in most animal cells tends to prevent _____
    charged ions from entering the cell by passive transport.

D. The membrane potential has no effect on the transport of _____ solutes.

concentration gradient; electrical potential difference; electrochemical gradient; membrane
potential; negatively; positively; uncharged.

12–5   Easy, multiple choice

The membrane potential of a typical animal cell favors the inward flux of:

A. water.
B. glucose.
C. Cl^-.
D. ATP.
E. Ca^{2+}.

Active Transport Moves Solutes Against Their Electrochemical Gradients (Pages 377–378)

12–6   Intermediate, short answer

Name the three main types of active transport.

12–7   Easy, multiple choice

The bacterial protein bacteriorhodopsin is an example of:

A. a light-driven pump.
B. a proton pump.
C. a passive transporter.
D. a coupled transporter.
E. a transmembrane protein.

Animal Cells Use the Energy of ATP Hydrolysis to Pump Out Na$^+$ (Pages 378–379)

12–8   Intermediate, multiple choice

If the plasma membrane of animal cells was made permeable to Na$^+$ and K$^+$, the Na$^+$-K$^+$ pump would:

A. be completely inhibited.

B. begin to pump Na$^+$ in both directions.

C. begin synthesizing ATP instead of hydrolyzing it.

D. continue to pump ions and to hydrolyze ATP but the energy of hydrolysis would be wasted, as it would generate heat rather than ion gradients.

E. continue to pump ions but would not hydrolyze ATP.

The Na$^+$-K$^+$ Pump Is Driven by the Transient Addition of a Phosphate Group (Pages 379–380)

12–9   Difficult, multiple choice + data interpretation

The Aeroschmidt weed contains an ATP-driven ion pump in its vacuolar membrane that pumps potentially toxic heavy metal ions such as Zn^{2+} and Pb^{2+} into the vacuole. The pump protein exists in a phosphorylated and an unphosphorylated form and works in a similar way to the Na$^+$-K$^+$ pump of animal cells. To study its action, you incorporate the unphosphorylated form of the protein into phospholipid vesicles containing K$^+$ in their interiors. (You ensure that all of the protein molecules are in the same orientation in the lipid bilayer.) When you add Zn^{2+} and ATP to the solution outside such vesicles, you find that Zn^{2+} is pumped into the vesicle lumen. You then expose vesicles containing the pump protein to the solutes shown in Table Q12–9A and determine the phosphorylation state of the protein in each sample. You get the results shown in Table Q12–9B.

Table Q12–9A

	A	B	C	D	E	F
Outside	Zn^{2+} + ATP	Zn^{2+}	Zn^{2+} + ATP	Zn^{2+}	ATP	ATP
Inside	K$^+$			K$^+$		K$^+$

Table Q12–9B

	A	B	C	D	E	F
Phosphorylated protein present	+	–	–	–	–	++
Unphosphorylated protein present	++	++	++	++	++	–

What would you expect to happen if you treat vesicles as in lane F, but before determining the phosphorylation state of the protein, you wash away the outside buffer and replace it with a buffer containing only Zn^{2+}?

A. Nothing will happen. (No Zn^{2+} will move into the vesicle; no K$^+$ will move out of the vesicle, the phosphorylation state of the protein will not change.)

B. No Zn^{2+} will move into the vesicle; no K$^+$ will move out of the vesicle; the protein will become unphosphorylated.

C. A small amount of Zn^{2+} will move into the vesicle; no K$^+$ will move out of the vesicle; the phosphorylation state of the protein will not change.

D. A small amount of Zn^{2+} will move into the vesicle; no K^+ will move out of the vesicle; the protein will become unphosphorylated.

E. A small amount of Zn^{2+} will move into the vesicle; a small amount of K^+ will move out of the vesicle; the phosphorylation state of the protein will not change.

Animal Cells Use the Na^+ Gradient to Take Up Nutrients Actively (Pages 380–381)

12–10   Easy, short answer

Explain why Na^+ is commonly used to drive the coupled inward transport of nutrients in animal cells.

12–11   Intermediate, multiple choice

Ouabain inhibits the active uptake of glucose into epithelial cells by:

A. binding to the glucose-Na^+ symport.

B. opening K^+ channels.

C. changing the pH of the cell.

D. increasing the intracellular concentration of Na^+.

E. depleting the cell of ATP.

The Na^+-K^+ Pump Helps Maintain the Osmotic Balance of Animal Cells (Pages 381–383)

12–12   Intermediate, multiple choice

Which of the following statements are true?

A. Amoebae have carrier proteins that actively pump water molecules from the cytoplasm to the cell exterior.

B. Bacteria and animal cells rely on the Na^+-K^+ pump in the plasma membrane to prevent lysis due to osmotic imbalances.

C. The Na^+-K^+ pump allows animal cells to thrive under conditions of very low ionic strength.

D. The cell wall surrounding plant cells prevents osmosis.

E. The Na^+-K^+ pump helps to keep both Na^+ and Cl^- ions out of the cell.

Intracellular Ca^{2+} Concentrations Are Kept Low by Ca^{2+} Pumps (Pages 383–384)

12–13   Easy, multiple choice

Ca^{2+} pumps in the plasma membrane and endoplasmic reticulum are important for:

A. maintaining osmotic balance.

B. preventing Ca^{2+} from altering the behavior of molecules in the cytosol.

    C. providing enzymes in the endoplasmic reticulum with Ca^{2+} ions that are necessary for their catalytic activity.

    D. maintaining a negative membrane potential.

    E. helping cells import K^+.

12–14   Easy, multiple choice

The Ca^{2+} pumps in the plasma membrane and endoplasmic reticulum are examples of:

    A. ATP-driven pumps.

    B. coupled transporters.

    C. passive carrier proteins.

    D. uniports.

    E. symports.

H^+ Gradients Are Used to Drive Membrane Transport in Plants, Fungi, and Bacteria (Pages 384–385)

12–15   Intermediate, multiple choice

Your first task as chief scientist at the Bubblehead Brewing Company is to purify an enzyme involved in fermentation in yeast. You grow some yeast cells at pH 7, resuspend them in a pH 7 buffer that contains a relatively high concentration of salt (NaCl), and lyse the cells. To your dismay, you find that the extract has no enzymatic activity. You measure the pH and discover that is has dropped from 7 to 5. Which of the following is the most reasonable explanation for the pH drop?

    A. The high salt concentration indirectly inhibited H^+ symport by inhibiting the Na^+-K^+ pump in the plasma membrane.

    B. The cells stopped metabolizing when they were transferred to the buffer, ATP was depleted, and the H^+ ATPase was unable to continue pumping protons.

    C. Disruption of the plasma membrane released H^+ ions from the cytoplasm.

    D. Disruption of the vacuolar membrane released H^+ ions from the vacuole.

ION CHANNELS AND THE MEMBRANE POTENTIAL (Pages 385–394)
Ion Channels Are Ion Selective and Gated (Pages 386–388)

12–16   Intermediate, multiple choice

Ion channels:

    A. only open in response to a signal of some kind.

    B. require input of energy in order to function.

    C. have no limit to the rate at which they can transport ions.

    D. can transport both negative and positive ions through the same channel.

    E. allow passage of ions in both directions.

Ion Channels Randomly Snap Between Open and Closed States (Pages 388–390)

12–17   Easy, multiple choice

A gated ion channel:

    A. stays continuously open when stimulated.

    B. opens more frequently in response to a given stimulus.

    C. opens more widely the stronger the stimulus.

    D. remains closed if unstimulated.

12–18   Intermediate, data interpretation

You have patch-clamped a single voltage-gated ion channel in a membrane and have obtained the recording shown in Figure Q12–18A.

(A) If you set the membrane potential equal to the threshold membrane potential this channel responds to, which of the recordings shown in Figure Q12–18B do you expect to observe? Explain why.

(B) What channel behavior does each of the other recordings show?

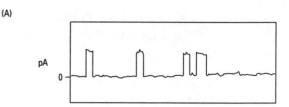

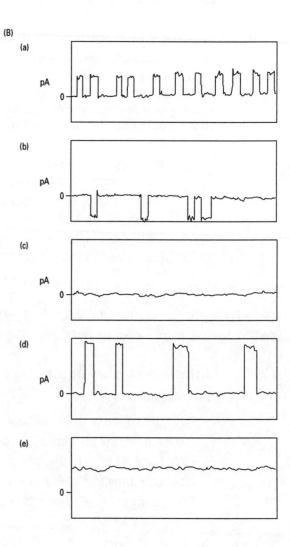

Q12–18

Voltage-gated Ion Channels Respond to the Membrane Potential (Pages 390–391)

12–19 Easy, matching/fill in blanks

For each of the following sentences, fill in the blank with the appropriate type of gated ion channel. You can use the same channel more than once.

A. The acetylcholine receptor in skeletal muscle cells is a _____ ion channel.

B. _____ ion channels are found in the hair cells of the mammalian cochlea.

C. _____ ion channels in the mimosa plant propagate the leaf-closing response.

D. _____ ion channels respond to changes in membrane potential.

E. Many receptors for neurotransmitters are _____ ion channels.

12–20 Intermediate/difficult, multiple choice + short answer

Which of the following statements are true? Explain your answer.

A. Voltage-gated channels are found only in nerve cells.

B. Voltage sensors are proteins that bind to voltage-gated ion channels and switch them to either their open or closed conformation.

C. K^+ leak channels are voltage-gated channels that help cells repolarize after an action potential has passed.

D. Ligand-gated channels differ from voltage-gated channels in that the binding of ligand opens the ligand-gated channel wider in addition to changing the frequency with which the channel opens.

E. The value of the threshold membrane potential for a voltage-gated channel depends on pH.

The Membrane Potential Is Governed by Membrane Permeability to Specific Ions (Pages 391–394)

12–21 Intermediate, multiple choice

The high intracellular concentration of K^+ in a resting animal cell is partly due to:

A. the K^+ leak channels in the plasma membrane.

B. the Na^+-K^+ pump in the plasma membrane.

C. voltage-gated K^+ channels in the plasma membrane.

D. intracellular stores of K^+ in the endoplasmic reticulum.

E. the membrane potential.

ION CHANNELS AND SIGNALNG IN NERVE CELLS (Pages 394–403)
Action Potentials Provide for Rapid Long-Distance Communication (Page 395)
Action Potentials Are Usually Mediated by Voltatge-gated Na⁺ Channels (Pages 395–397)

12–22   Easy, matching/fill in blanks

For each of the following sentences, fill in the blanks with the correct word or phrase selected from the list below. Use each word or phrase only once.

A. The action potential is a wave of _____ that rapidly spreads along the neuronal plasma membrane. This wave is triggered by a local change in the membrane potential to a value that is _____ negative than the resting membrane potential.

B. The action potential is propagated by the opening of _____-gated channels.

C. During an action potential, the membrane potential changes from _____ to _____.

D. The action potential travels from the neuron's _____ along the _____ to the nerve terminals.

E. Neurons chiefly receive signals at their highly branched _____.

hyperpolarization; more; depolarization; anions; negative; positive; cell body; axon; pressure; dendrites; neutral; less; cytoskeleton; ligand; voltage.

12–23   Easy, data interpretation

On the diagram of an action potential in Figure Q12–23, draw a dashed line showing what would have happened to the membrane potential after the initial depolarizing stimulus if there had been no voltage-gated Na⁺ channels in the membrane.

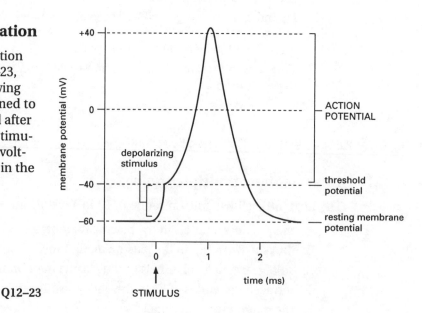

Q12–23

12–24   Intermediate, multiple choice

The action potential travels in one direction because:

A. the Na⁺-K⁺ pump restores the concentrations of Na⁺ and K⁺ to their original levels.

B. the K⁺ leak channels allow K⁺ to flow out, restoring the membrane to the resting potential.

C. depolarization of the membrane causes voltage-gated K^+ channels to open.

D. voltage-gated Na^+ channels adopt a transitory inactive conformation after being opened.

E. voltage-gated Na^+ channels spend less time in the open conformation when the membrane returns to the resting potential.

Voltage-gated Ca^{2+} Channels Convert Electrical Signals into Chemical Signals at Nerve Terminals (Pages 397–399)

Transmitter-gated Channels in Target Cells Convert Chemical Signals Back into Electrical Signals (Pages 399–400)

12–25   Easy, matching/fill in blanks

For each of the following sentences, fill in the blanks with the correct word or phrase selected from the list below. Use each word or phrase only once.

A. Neurons communicate with each other through specialized sites called _____.

B. Many neurotransmitter receptors are ligand-gated ion channels that open transiently in the _____ cell membrane in response to neurotransmitters released by the _____ cell.

C. The _____ is a ligand-gated Cl^- ion channel in the plasma membrane of _____ cells.

D. Ligand-gated ion channels in nerve cell membranes convert _____ signals into _____ ones.

E. Neurotransmitter release is stimulated by the opening of voltage-gated _____ in the nerve terminal membrane.

postsynaptic; Na^+ channels; chemical; K^+ channels; acetylcholine receptor; presynaptic; electrical; synapses; muscle; nerve; Ca^{2+} channels; GABA receptor.

12–26   Intermediate/difficult, multiple choice

You place two nerve cells, which have synaptic connections to each other and are normally able to signal to each other, in a dish filled with a buffer containing H^+, Na^+, and Cl^- as the only inorganic ions. You stimulate the first neuron by applying a depolarizing voltage across the membrane of one of the dendrites but find that the second neuron does not respond. Which of the following is the most likely explanation for its lack of response?

A. An action potential can only be initiated by opening a ligand-gated channel.

B. Propagation of the action potential along the axon is prevented by the lack of K^+ in the buffer.

C. Neurotransmitter is not being released into the synaptic cleft.

D. The axon membrane is unable to repolarize in this buffer.

E. Neurotransmitter is released but is unable to stimulate opening of the ligand-gated channels in the second neuron.

Neurons Receive Both Excitatory and Inhibitory Inputs (Pages 400–401)

12–27   Intermediate, multiple choice

Which of the following statements are true?

- A. Excitatory neurotransmitters cause the membrane potential to become more negative.
- B. The neurotransmitter GABA acts by making cells harder to depolarize.
- C. Acetylcholine and glycine are excitatory neurotransmitters.
- D. Inhibitory neurotransmitters act by stimulating the outflow of Cl^- from the cell.
- E. Tranquilizers such as Valium work by inhibiting the action of neurotransmitter receptors.

Synaptic Connections Enable You to Think, Act, and Remember (Pages 401–403)

12–28   Easy, multiple choice

Which of the following is NOT a benefit of using a chain of neurons and chemical synapses rather than a direct connection between the site of stimulus and the site of response?

- A. Each chemical synapse represents an opportunity for the organism to modify the signal being sent.
- B. Diffusion of small molecules is more rapid than propagation of an electric signal; thus the signal is speeded up by having more synapses.
- C. Use of chemical synapses increases the variety of messages the presynaptic cell can send to the postsynaptic cell.
- D. Modification of different ion channels at synapses can be used to generate memory on the cellular level.
- E. Chemical synapses allow neurons to receive and integrate input from more than one source.

Answers

A12–1. D.

A12–2. A. Carrier proteins and channel proteins can provide a <u>hydrophilic</u> pathway through the membrane for specific polar solutes or inorganic ions.

B. A substance is transported down its concentration gradient by <u>passive</u> transport

C. A substance is transported up its concentration gradient by <u>active</u> transport.

D. <u>Channel</u> proteins can transport a substance down its concentration gradient only.

E. <u>Carrier</u> proteins are highly selective in the solutes they transport, binding the solute at a specific site.

A12–3. C. Movement of Na^+ out of cells requires an input of energy and therefore does not occur passively through a channel. Amino acid and glucose transport occurs through carrier proteins, not channels.

A12–4. A. The net force driving a charged solute across a cell membrane is the <u>electrochemical gradient.</u> This is composed of two forces, one due to the <u>concentration gradient</u> of the solute and the other due to the <u>electrical potential difference (voltage)</u> across the membrane, which is known as the <u>membrane potential</u>.

B. In most circumstances the inside of the cell is <u>negatively</u> charged with respect to the outside.

C. The membrane potential in most animal cells tends to prevent <u>negatively</u> charged ions from entering the cell by passive transport.

D. The membrane potential has no effect on the transport of <u>uncharged</u> solutes.

A12–5. E. The membrane potential is negative and therefore favors the inward flux of positive ions only. Uncharged molecules like water and glucose are not affected by the membrane potential.

A12–6. 1, ATP-driven transport. 2, coupled transport. 3, light-driven transport.

A12–7. A, B, and E.

A12–8. D. If the membrane became permeable to Na^+ and K^+, the concentrations of these ions would tend to become equal on both sides of the membrane. The pump would still be able to function (i.e., it would still hydrolyze ATP and pump K^+ in and Na^+ out), but the transported ions would then move back down their concentration gradients, releasing heat in the process.

A12–9. D. If the pump is mechanistically similar to the Na^+-K^+ pump, then the transport of ions is driven by ATP hydrolysis, the pump is transiently phosphorylated, phosphorylation is stimulated by one ion and dephosphorylation is stimulated by the other ion. Since all of the protein is in the phosphorylated form in the absence of Zn^{2+} (lane F), Zn^{2+} is probably required for dephosphorylation. K^+, then, probably binds to the dephosphorylated form and stimulates the ATPase/autophosphorylation. So, if Zn^{2+} is added to the phosphorylated pump, Zn^{2+} will stimulate dephosphorylation, trigger a conformational change, and be injected into the vesicle. K^+ will stimulate the kinase activity of the pump, but since there is no ATP to be hydrolyzed in the interior of the vesicle, no phosphorylation and hence no movement of K^+ will occur.

A12–10. Na^+ is commonly used to drive coupled transport in animal cells because a steep concentration gradient of Na^+ (high outside and low inside) is maintained by the Na^+-K^+ pump. Na^+ readily flows back into the cell down this gradient because of the negative membrane potential. The energy provided by the flow of Na$^+$ down this steep electrochemical gradient can be harnessed by coupled transporters.

A12–11.   D. Ouabain inhibits the Na$^+$-K$^+$ pump and therefore leads to an increase in the intracellular concentration of Na$^+$, which is continually leaking into the cell. Uptake of glucose into epithelial cells occurs via an Na$^+$-glucose symport, which uses the Na$^+$ gradient to drive movement of glucose into the cell.

A12–12.   E. The Na$^+$-K$^+$ pump keeps Na$^+$ out directly by pumping it out and keeps Cl$^-$ out indirectly by helping to maintain the negative membrane potential. Cells do not have pumps for moving water molecules across the membrane (A), since the lipid bilayer is permeable to water. Bacteria do not have Na$^+$-K$^+$ pumps in their plasma membranes (B). The Na$^+$-K$^+$ pump cannot directly remove water molecules from the cell; it helps maintain osmotic balance by pumping out the Na$^+$ that leaks in, which would not help if the cell is in a solution of very low ionic strength (C). The plant cell wall is permeable to water and therefore cannot prevent osmosis (D).

A12–13.   B. The major purpose of the Ca^{2+} pumps is to keep the cytosolic concentration of Ca^{2+} low. When Ca^{2+} does move into the cytosol, it alters the behavior of many proteins; hence Ca^{2+} is a powerful signaling molecule. It is not involved in the catalytic activity of ER enzymes (C). Since the levels of Ca^{2+} are very low relative to the levels of K$^+$ and Na$^+$, the Ca^{2+} gradient does not have a significant effect on the osmotic balance of the cell (A) or the membrane potential (D). It is not involved in K$^+$ import (E).

A12–14.   A and D.

A12–15.   D. Yeast do not have a Na$^+$-K$^+$ pump in the plasma membrane. The H$^+$ ATPase in the plasma membrane pumps H$^+$ ions out of the cell; hence, inhibition of the pump would—if anything— lead to an increase in the pH (and lysis of the cells afterwards would ultimately release these ions back into the buffer). The plasma membrane H$^+$ ATPase causes the cytoplasm of yeast cells to be alkaline relative to the medium that they are growing in, so lysis of the plasma membrane alone would not cause the pH of the buffer to drop. However, the vacuole contains a high concentration of H$^+$ ions as a result of the action of the vacuolar membrane H$^+$ ATPase; these are sequestered from the buffer and the growth medium until the cells are lysed.

A12–16.   E. Ions can pass either way through a channel; the direction in which flow takes place depends on the concentration gradient and the membrane potential. Some ion channels require a specific stimulus to open them; others, such as K$^+$ leak channels, do not (A). Ion channels are passive transporters and therefore require no energy source in order to function (B). Ion channels are very fast relative to carrier proteins but are limited by the rate at which ions can move through the channel (C). An ion channel allows specific positive or negative ions to pass, but not both (D).

A12–17.   B.

A12–18.   (A) Recording (a). It shows a channel that is spending more time in the open conformation, as one would expect for a voltage-gated channel that has been subjected to its threshold membrane potential. (B). Recording (b) shows a channel where the ion flow is reversed compared to recording (a). Recording (c) shows a channel that is closed all of the time. Recording (d) shows a channel that is allowing twice as much current to pass through as the original channel. This must be a different channel, as a given channel cannot let more current through by opening more widely. Recording (e) shows a channel that is open all of the time.

A12–19.   A. The acetylcholine receptor in skeletal muscle cells is a <u>ligand-gated</u> ion channel.

B. <u>Stress-activated</u> ion channels are found in the hair cells of the mammalian cochlea.

C. <u>Voltage-gated</u> ion channels in the mimosa plant propagate the leaf-closing response.

D. <u>Voltage-gated</u> ion channels respond to changes in membrane potential.

E. Many receptors for neurotransmitters are <u>ligand-gated</u> ion channels.

A12–20.   E. The threshold membrane potential at which a voltage-gated channel changes conformation depends on the charged residues in the voltage sensor domain. Since the charge on an amino acid side chain depends on the pH of the surroundings, the threshold potential must also be sensitive to pH. Voltage-gated channels are found in a number of different types of cells, including plant cells (A). Voltage sensors are special voltage-sensitive domains of voltage-gated channels (B). K^+ leak channels are not voltage-gated (C). The width of ligand-gated channels is not affected by ligand binding (D).

A12–21.   B and E. The Na^+-K^+ pump continually transports K^+ into the cell. The negative membrane potential also helps to retain K^+ in the cell. The K^+ leak channels allow K^+ to move both into and out of the cell and so do not contribute to the accumulation of K^+ in the cell.

A12–22.   A. The action potential is a wave of <u>depolarization</u> that rapidly spreads along the neuronal plasma membrane. This wave is triggered by a local change in the membrane potential to a value that is <u>less</u> negative than the resting membrane potential.

B. The action potential is propagated by the opening of <u>voltage</u>-gated channels.

C. During an action potential, the membrane potential changes from <u>negative</u> to positive.

D. The action potential travels from the neuron's <u>cell body</u> along the <u>axon</u> to the nerve terminals.

E. Neurons chiefly receive signals at the highly branched <u>dendrites</u>.

A12–23.   Figure A12–23.

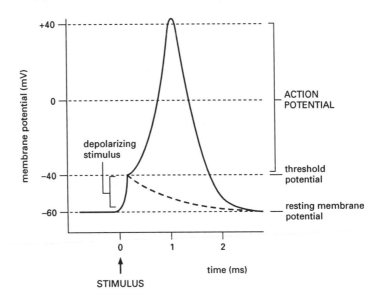

A12–23

A12–24.   D. The temporary inactive conformation of the voltage-gated Na^+ channels prevents the action potential from moving back toward previously stimulated patches of the membrane, as well as causing these portions of the membrane to be temporarily refractory to a new action potential. Statements A–C, and E explain how the membrane returns to the resting potential after the action potential has passed.

A12–25.   A. Neurons communicate with each other through specialized sites called <u>synapses</u>.

B. Many neurotransmitter receptors are ligand-gated ion channels that open transiently in the <u>postsynaptic</u> cell membrane in response to neurotransmitters released by the <u>presynaptic</u> cell.

C. The <u>GABA receptor</u> is a ligand-gated Cl^- ion channel in the plasma membrane of <u>nerve</u> cells.

D. Ligand-gated ion channels in nerve cell membranes convert <u>chemical</u> signals into <u>electrical</u> ones.

E. Neurotransmitter release is stimulated by the opening of voltage-gated <u>Ca^{2+} channels</u> in the nerve terminal membrane.

A12–26.   C. Since there is no Ca^{2+} in the buffer, when the action potential reaches the synapse and causes Ca^{2+} channels to open, no Ca^{2+} will flow in, and neurotransmitter will not be released. Ligand-gated channels trigger the action potential by causing a temporary depolarization of the membrane; therefore applying a depolarizing voltage across the membrane will also trigger an action potential (A). Na^+ in the buffer (not K^+) is required for propagation of the action potential (B). Repolarization is due to leakage of K^+ out of cells and should therefore occur under the described conditions (D). (Also, repolarization is not necessary for the initial propagation of the action potential along the axon.) E is not impossible, but unlikely.

A12–27.   B. GABA is an inhibitory neurotransmitter and opens Cl^- channels in the membrane, allowing an inflow of Cl^-, which makes cells harder to depolarize. Excitatory neurotransmitters cause the membrane to become less negative, thereby making it easier for the membrane to be depolarized to the threshold potential (A). Glycine is an inhibitory neurotransmitter (C). Inhibitory neurotransmitters that act by affecting Cl^- ion movement tend to stimulate the inflow of Cl^-, not the outflow. (D). Valium acts by binding to GABA-gated Cl^- channels, making them easier to open (E).

A12–28.   B. Diffusion of small molecules is slower than the movement of an electrical signal, and having more synapses increases the total time it takes to deliver a message.

13 Energy Generation in Mitochondria and Chloroplasts

Questions

Cells Obtain Most of Their Energy by a Membrane-based Mechanism (Pages 409–410)

13–1   Easy, multiple choice

Which of the following statements are most likely to be true?

    A. Organisms that could carry out fermentation evolved before those that could carry out aerobic respiration.

    B. Organisms that could carry out oxygen-producing photosynthesis evolved before those that could carry out fermentation reactions.

    C. Eucaryotic organisms were the first to evolve mechanisms of chemiosmotic coupling.

    D. Both photosynthesis and aerobic respiration first evolved in procaryotes.

    E. Aerobic respiration arose as an adaptation to increasing levels of oxygen in the atmosphere that had been produced by photosynthesis.

13–2   Easy, short answer

What three essential items are missing from the following list of cellular components required to make ATP by chemiosmotic coupling?

ADP; ATP synthase; protons; electron-transport chain; proton pump; membrane.

13–3   Intermediate, multiple choice (Requires student to have studied the whole chapter)

Electron transport is coupled to ATP synthesis in mitochondria, chloroplasts, and thermophilic bacteria such as *Methanococcus*. Which of the following are likely to affect the coupling of electron transport to ATP synthesis in ALL of these systems?

    A. A potent inhibitor of cytochrome oxidase.

    B. The removal of oxygen.

    C. The absence of light.

    D. An ADP analogue that inhibits ATP synthase.

    E. Dinitrophenol (permeabilizes membranes to protons).

MITOCHONDRIA AND OXIDATIVE PHOSPHORYLATION (Pages 410–421)
A Mitochondrion Contains Two Membrane-bounded Compartments (Pages 411–413)

13–4   Easy, short answer

In which of the four compartments of a mitochondrion are each of the following located?

    A. Porin.

    B. The mitochondrial genome.

    C. Citric acid cycle enzymes.

    D. Proteins of the electron transport chain.

    E. ATP synthase.

    F. Membrane transport protein for pyruvate.

13–5   Easy, multiple choice (Requires information from section on pages 417–419)

Which of the following statements about mitochondria are false?

    A. Protons are pumped from the intermembrane space into the matrix.

    B. ATP is synthesized in the matrix.

    C. Mitochondria can change shape.

    D. The outer membrane is permeable to protons.

    E. The inner membrane is folded into cristae.

High-Energy Electrons Are Generated via the Citric Acid Cycle (Pages 413–414)

13–6   Easy, matching/fill in blanks

For each of the following sentences, fill in the blanks with the correct word or phrase selected from the list below. Use each word or phrase only once.

    A. Mitochondria can use both _____ and _____ as fuel.

    B. _____ produced in the citric acid cycle donates electrons to the electron-transport chain.

    C. The citric acid cycle oxidizes _____ and produces _____ as a waste product.

    D. _____ acts as the final electron acceptor in the electron-transport chain.

    E. The synthesis of ATP in mitochondria is also known as _____.

NADH; oxygen; pyruvate; oxidative phosphorylation; fatty acids; glucose; NADPH; NAD^+; $NADP^+$; acetyl groups; carbon dioxide; chemiosmosis.

Electrons Are Transferred Along a Chain of Proteins in the Inner Mitochondrial Membrane (Pages 414–415)

13–7   Easy, matching/fill in blanks

For each of the following sentences, fill in the blanks with the correct word or phrase selected from the list below. Use each word or phrase only once.

    A. NADH donates electrons to the _____ of the three respiratory enzyme complexes in the mitochondrial electron-transport chain.

    B. _____ is a small protein that acts as a mobile electron carrier in the respiratory chain.

    C. _____ transfers electrons to oxygen.

    D. Electron transfer in the chain occurs in a series of _____ reactions.

    E. The first mobile electron carrier in the respiratory chain is _____.

NADH dehydrogenase; first; the cytochrome b-c_1 complex; ubiquinone; plastoquinone; second; third; oxidation; cytochrome c; cytochrome oxidase; oxidation-reduction; phosphorylation; reduction.

13–8   Intermediate, multiple choice

Which of the following explains why adding reduced ubiquinone to mitochondria lacking cytochrome c does not lead to production of a proton gradient?

    A. Formation of the proton gradient requires oxidized, not reduced, ubiquinone.

    B. Ubiquinone is too weak a reducing agent to donate electrons to cytochrome oxidase.

    C. In the absence of cytochrome c, the cytochrome b-c_1 complex passes electrons directly to oxygen.

    D. Ubiquinone cannot bind to cytochrome oxidase as required to pass electrons to it.

    E. Cytochrome oxidase reduces O_2 but does not pump protons.

Electron Transport Generates a Proton Gradient Across the Membrane (Pages 415–417)

13–9   Intermediate, multiple choice

Which of the following statements are true?

    A. Because the electrons in NADH are at a higher energy than the electrons in reduced ubiquinone, the NADH dehydrogenase complex can pump more protons than can the cytochrome b-c_1 complex.

    B. The pH in the mitochondrial matrix is higher than the pH in the intermembrane space.

    C. The proton concentration gradient and the membrane potential across the inner mitochondrial membrane tend to work against each other in driving protons from the intermembrane space into the matrix.

D. The difference in proton concentration across the inner mitochondrial membrane has a much larger effect on the total proton-motive force that does the membrane potential.

E. All of the free energy released by the net transfer of electrons from NADH to O_2 is captured in the form of the proton gradient.

The Proton Gradient Drives ATP Synthesis (Pages 417–419)

13–10   Intermediate, short answer

Some bacteria can live both aerobically and anaerobically. How does the ATP synthase in the plasma membrane of the bacterium enable such bacteria to keep functioning in the absence of oxygen?

13–11   Intermediate/difficult, multiple choice

Bongkrekic acid is an antibiotic that inhibits the ATP/ADP transport protein in the inner mitochondrial membrane. Which of the following will allow electron transport to occur in mitochondria treated with bongkrekic acid?

A. Placing the mitochondria in anaerobic conditions.
B. Adding $FADH_2$.
C. Permeabilizing the inner membrane to protons.
D. Inhibiting the ATP synthase.
E. Increasing the concentration of ATP in the matrix.

Coupled Transport Across the Inner Mitochondrial Membrane Is Driven by the Electrochemical Proton Gradient (Page 419)

13–12   Intermediate, multiple choice

Which of the following types of ion movement might be expected to require cotransport of protons from the intermembrane space to the matrix, inasmuch as it could not be driven by the membrane potential across the inner membrane? (Assume that each ion being moved is moving against its concentration gradient.)

A. Import of Ca^{2+} into the matrix from the intermembrane space.
B. Import of acetate ion into the matrix from the intermembrane space.
C. Exchange of Fe^{2+} in the matrix for Fe^{3+} in the intermembrane space.
D. Export of Cl^- from the matrix to the intermembrane space.
E. Exchange of Ca^{2+} in the matrix for Na^+ in the intermembrane space.

Proton Gradients Produce Most of the Cell's ATP (Pages 419–421)

13–13   Intermediate, short answer

How many molecules of ATP are produced by the electron-transport chain and the ATP synthase from the oxidation of the 18-carbon fatty acid derivative stearyl CoA? The first steps in the oxidation of stearyl CoA occur in the mitochondrion and generate 9 molecules of acetyl CoA, 8 molecules of NADH, and 8 molecules of $FADH_2$.

ELECTRON-TRANSPORT CHAINS AND PROTON PUMPING (Pages 421–430)

13–14   Intermediate, multiple choice

Generation of a proton gradient by electron-transport proteins:

    A. is possible because high-energy electrons have a higher affinity for protons than do low-energy electrons.
    B. requires that the oxidized and reduced states of the electron-transport protein have different conformations.
    C. occurs by a mechanism that allows only one proton to be pumped per electron that passes through a complex.
    D. requires a reducing agent like NADH that can donate both electrons and protons to the first carrier in the chain.

The Redox Potential Is a Measure of Electron Affinities (Pages 422–423)

13–15   Intermediate, multiple choice

Which of the following statements are true?

    A. Only compounds with negative redox potentials can donate electrons to other compounds under standard conditions.
    B. Compounds that donate one electron have higher redox potentials than those of compounds that donate two electrons.
    C. The $\Delta E_0'$ of a redox pair does not depend on the concentration of each member of the pair.
    D. The free energy change, ΔG, for an electron transfer reaction does not depend on the concentration of each member of a redox pair.
    E. If the E_0' of the reaction $AH_2 \rightarrow A + 2\,H^+ + 2e^-$ is 600 mV and the E_0' of the reaction $H_2O \rightarrow 1/2\,O_2 + 2\,H^+ + 2e^-$ is 820 mV, then the transfer of electrons from AH_2 to water must be favorable under standard conditions.

Electron Transfers Release Large Amounts of Energy (Pages 423–425)

13–16   Intermediate, data interpretation

With reference to the table below, which of the following reactions have a large enough free energy change to enable it to be used, in principle, to provide the energy needed to synthesize one molecule of ATP from ADP and P_i under standard conditions?

    A. The reduction of a molecule of pyruvate by NADH.

B. The reduction of a molecule of cytochrome b by NADH.

C. The reduction of a molecule of cytochrome b by reduced ubiquinone.

D. The oxidation of a molecule of reduced ubiquinone by cytochrome c.

E. The oxidation of cytochrome c by oxygen.

Reaction	E'_0
$NADH \rightarrow NAD^+ + H^+ + 2e^-$	-320 mV
Lactate $\rightarrow$ pyruvate $+ 2H^+ + 2e^-$	-190 mV
Reduced ubiquinone $\rightarrow$ ubiquinone $+ 2H^+ + 2e^-$	30 mV
Cytochrome b $(Fe^{2+}) \rightarrow$ cytochrome b $(Fe^{3+}) + e^-$	70 mV
Cytochrome c $(Fe^{2+}) \rightarrow$ cytochrome c $(Fe^{3+}) + e^-$	230 mV
$H_2O \rightarrow 1/2 O_2 + 2H^+ + 2e^-$	820 mV

Metals Tightly Bound to Proteins Form Versatile Electron Carriers (Pages 425–427)

13–17   Intermediate/difficult, multiple choice

Which of the following statements are true?

A. Ubiquinone is a small hydrophobic protein containing a metal group that acts as an electron carrier.

B. A 2Fe2S iron-sulfur center carries one electron while a 4Fe4S center carries two.

C. Iron-sulfur centers generally have a higher redox potential than do cytochromes.

D. Mutation of hydrophobic amino acids near the heme group of cytochrome c to acidic amino acids is likely to increase the redox potential of cytochrome c.

E. Mitochondrial electron carriers with the highest redox potential generally contain copper ions and/or heme groups.

Protons Are Pumped Across the Membrane by the Three Respiratory Enzyme Complexes (Pages 427–429)
Respiration Is Amazingly Efficient (Pages 429–430)

13–18   Intermediate, short answer

When molecular oxygen (O_2) picks up one electron it becomes converted to the superoxide radical O_2^-. This radical is potentially damaging to cells as it will avidly pick up another three electrons from a wide variety of cellular molecules. How do cells avoid this happening during cellular respiration?

CHLOROPLASTS AND PHOTOSYNTHESIS (Pages 430–439)
Chloroplasts Resemble Mitochondria but Have an Extra Compartment (Pages 430–432)

13–19   Intermediate, multiple choice

Ions are able to diffuse freely:

A. from one thylakoid space to another thylakoid space in the same chloroplast.

    B. from the cytosol to the intermembrane space.

    C. from the intermembrane space to the stroma.

    D. from the stroma to the thylakoid space.

    E. from the interior of one granum to another in the same chloroplast.

Chloroplasts Capture Energy from Sunlight and Use It to Fix Carbon (Pages 432–433)

13–20   Easy, short answer

Write out the substrates and products for the net reactions that occur in the "light" and "dark" stages of photosynthesis.

Excited Chlorophyll Molecules Funnel Energy into a Reaction Center (Pages 433–434)

13–21   Intermediate, multiple choice + data interpretation

If you shine light on chloroplasts and measure the rate of photosynthesis as a function of light intensity, you get a curve that plateaus at a fixed rate of photosynthesis, x, as shown in Figure Q13–21.

Which of the following conditions will increase the value of x?

    A. Increasing the number of chlorophyll molecules in the antennae complexes.

    B. Increasing the number of reaction centers.

    C. Adding a powerful oxidizing agent.

    D. Decreasing the wavelength of light used.

    E. Increasing the rate at which chlorophyll molecules are able to transfer electrons to one another.

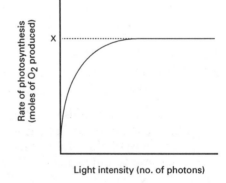

Q13–21

Light Energy Drives the Synthesis of ATP and NADPH (Pages 434–436)

13–22   Difficult, multiple choice

If you add a compound to illuminated chloroplasts that inhibits the NADP reductase, NADPH generation ceases, as expected. Ferredoxin, however, does not accumulate in the reduced form because it is able to donate its electrons not only to NADP (via NADP reductase), but also back to the cytochrome b_6-f complex. Thus, in the presence of the above compound, a "cyclic" form of photosynthesis occurs in which electrons flow in a circle from ferredoxin, to the cytochrome b_6-f complex, to plastocyanin, to photosystem I, to ferredoxin. What will happen if you now also inhibit photosystem II?

    A. Less ATP will be generated per photon absorbed.

    B. ATP synthesis will cease.

    C. Plastoquinone will accumulate in the oxidized form.

D. Plastocyanin will accumulate in the oxidized form.

E. Ferredoxin will accumulate in the reduced form.

Carbon Fixation Is Catalyzed by Ribulose Bisphosphate Carboxylase (Pages 436–438)
Carbon Fixation in Chloroplasts Generates Sucrose and Starch (Page 438)

13–23   Easy, matching/fill in blanks (Note to instructors: Questions 13–23 and 13–24 are alternatives and should not be used on the same test paper)

For each of the following sentences, fill in the blanks with the correct word or phrase selected from the list below. Use each word or phrase only once.

A. In the carbon fixation process in chloroplasts, carbon dioxide is initially added to the sugar _____.

B. The final product of carbon fixation in chloroplasts is the three-carbon compound _____.

C. This is converted into _____, which can be directly used by the mitochondria, into _____, which is exported to other cells, and into _____, which is stored in the stroma.

D. The carbon fixation cycle requires energy in the form of _____ and reducing power in the form of _____.

NADPH; NADH; pyruvate; glyceraldehyde 3-phosphate; 3-phosphoglycerate; starch; ATP; ribose 1,5-bisphosphate; sucrose; ribulose 1,5-bisphosphate.

13–24   Difficult, multiple choice (Note to instructors: Questions 13–24 and 13–23 are alternatives and should not be used on the same test paper)

The enzyme ribulose bisphosphate carboxylase (rubisco) normally adds carbon dioxide to ribulose 1,5-bisphosphate. However, it will also catalyze a competing reaction in which O_2 is added to ribulose 1,5-bisphosphate to form 3-phosphoglycerate and phosphoglycolate. Assume that phosphoglycolate is a compound that cannot be used in any further reactions. If O_2 and CO_2 have the same affinity for rubisco, which of the following is the lowest ratio of CO_2 to O_2 at which a net synthesis of sugar can occur?

A. 1:3.

B. 1:2.

C. 1:1.

D. 2:1.

E. 3:1.

OUR SINGLE-CELLED ANCESTORS (Pages 439–442)
RNA Sequences Reveal Evolutionary History (Pages 439–440)

13–25   Intermediate, multiple choice (Requires student to have studied Chapter 1)

Which of the phylogenetic trees in Figure Q13–25 is the most accurate? (The mitochondria and chloroplasts are from maize, but are treated as independent "organisms" for the purposes of this question.)

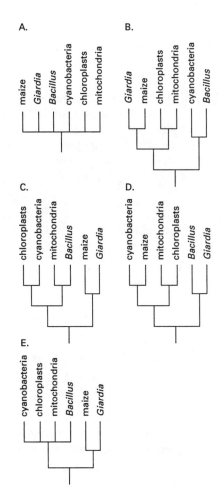

Q13–25

Ancient Cells Probably Arose in Hot Environments (Pages 440–441)
Methanococcus **Lives in the Dark, Using Only Inorganic Materials as Food (Pages 441–442)**

13–26   Intermediate, multiple choice

The biology of *Methanococcus jannaschii* is of interest to evolutionary biologists because:

A. phylogenetically it is more closely related to the proposed ancestor cell than are organisms such as cyanobacteria and yeast.

B. its small genome is approximately the same size as the genome of the proposed ancestor cell.

C. it lives under conditions that resemble the environment in which the first living cells may have dwelt.

D. its method of carbon fixation is thought to have given rise to the dark reactions used in modern cyanobacteria and chloroplasts.

E. the proteins in the electron-transport chain that *Methanococcus* uses to generate energy are thought to have given rise to those in the electron-transport chain used in modern mitochondria and aerobic bacteria.

Answers

A13–1. A, D, and E.

A13–2. inorganic phosphate (P_i); an electron donor that provides high-energy electrons (*or* high-energy electrons); an electron acceptor.

A13–3. D and E. All chemiosmotic coupling systems involve a proton gradient that is utilized by an ATP synthase that binds ADP and phosphorylates it. Hence all chemiosmotic systems will be affected by agents that prevent ADP from binding the synthase or that dissipate the proton gradient. Cytochrome oxidase and oxygen are required only for mitochondria and aerobic bacteria (not *Methanococcus*); light is required only for chloroplasts and photosynthetic bacteria (not *Methanococcus*).

A13–4. Porin = outer membrane; mitochondrial genome = matrix; citric acid cycle enzymes = matrix; proteins in the electron-transport chain = inner membrane; ATP synthase = inner membrane; transport protein for pyruvate = inner membrane.

A13–5. A.

A13–6. A. Mitochondria can use both <u>pyruvate</u> and <u>fatty acids</u> directly as fuel.

B. <u>NADH</u> produced in the citric acid cycle donates electrons to the electron-transport chain.

C. The citric acid cycle oxidizes <u>acetyl groups</u> and produces <u>carbon dioxide</u> as a waste product.

D. <u>Oxygen</u> acts as the final electron acceptor in the electron-transport chain.

E. The synthesis of ATP in mitochondria is also known as <u>oxidative phosphorylation</u>.

A13–7. A. NADH donates electrons to the <u>first</u> of the three respiratory enzyme complexes in the mitochondrial electron-transport chain.

B. <u>Cytochrome c</u> is a small protein that acts as a mobile electron carrier in the respiratory chain.

C. <u>Cytochrome oxidase</u> transfers electrons to oxygen.

D. Electron transfer in the chain occurs in a series of <u>oxidation-reduction</u> reactions.

E. The first mobile electron carrier in the respiratory chain is <u>ubiquinone</u>.

A13–8. D. Formation of a proton gradient can occur in mitochondria lacking cytochrome c if one provides an electron acceptor for the cytochrome b-c_1 complex or an electron donor for the cytochrome c oxidase. Thermodynamically, oxidized ubiquinone cannot accept electrons from the cytochrome b-c_1 complex, but it is possible for reduced ubiquinone to donate electrons to the cytochrome c oxidase complex, since reduced ubiquinone is a stronger reducing agent than cytochrome c and freely diffuses through the membrane. However, each of the three electron-transport complexes is specific as to which substrates they can bind and use, and since reduced ubiquinone does not bind to cytochrome oxidase, no electron transfer takes place.

A13–9. B. The pumping of protons out of the matrix into the intermembrane space creates a difference in proton concentration on both sides of the membrane, with the matrix at a higher pH (i.e., more alkaline) than the intermembrane space, which tends to equilibrate with the cytosol, which is at neutral pH. The electrons in NADH are at a higher energy than the electrons in reduced ubiquinone (A), but it is the difference in energy of the electrons in the two substrate/product pairs (i.e., NADH/reduced ubiquinone and reduced ubiquinone/cytochrome c (reduced)) that determines how many protons can be pumped by each complex. The proton concentration gradient and the membrane potential generated by the electron-transport chain work in the same direction (C), creating a steep electrochemical gradient for protons across the membrane. D, The difference in proton concentration has a smaller effect on the total proton-motive force than does the membrane potential. E, All of the free energy released by the net transfer of electrons from NADH to O_2 cannot be captured in the form of the proton gradient; some of the free energy used to drive the reactions is lost as heat.

A13–10.   In the absence of oxygen, the respiratory chain no longer pumps protons and thus no proton electrochemical gradient is generated across the bacterial membrane. In these conditions the ATP synthase uses some of the ATP generated by glycolysis in the cytosol to pump protons out of the bacterium, thus forming the proton gradient across the membrane that the bacterium requires in order to import vital nutrients by coupled transport.

A13–11.   C. Inhibition of the ATP/ADP translocase prevents export of ATP generated by oxidative phosphorylation in exchange for an import of ADP into the matrix. The ensuing buildup of ATP at the expense of ADP inhibits the ATP synthase. Since protons are no longer being used to power the ATP synthase, the proton gradient is not dissipated; it therefore becomes increasingly difficult for the electron-transport proteins to pump protons out of the matrix and electron transport quickly stops. Hence, permeabilizing the inner membrane to protons will allow electron transport to resume (although no ATP will be synthesized).

A13–12.   B, and E. Since the inside of the matrix is more negative than the outside, in principle, any traffic resulting in an increase in the positive charge in the matrix can be driven by the membrane potential. Hence, import of Ca^{2+} into the matrix, exchange of Fe^{2+} in the matrix for Fe^{3+} in the intermembrane space, and export of Cl^- from the matrix can be driven by the membrane potential and need not require the cotransport of protons down the pH gradient. Import of acetate ion into the matrix and exchange of Ca^{2+} in the matrix for Na^+ in the intermembrane space, in contrast, result in an increase in the amount of negative charge in the matrix and cannot be driven by the charge difference between the two mitochondrial compartments.

A13–13.   113. Since each molecule of acetyl CoA produces 3 molecules of NADH and 1 molecule of $FADH_2$, the oxidation of stearyl CoA generates a total of 35 molecules of NADH and 17 molecules of $FADH_2$. From each molecule of NADH and $FADH_2$, 2.5 and 1.5 molecules of ATP can be formed, respectively. In addition, 9 molecules of GTP are produced by the citric acid cycle, prior to electron transport.

A13–14.   B. The mechanism of proton pumping by electron-transport proteins involves the reduced form of the transport protein picking up protons from one side of the membrane and the oxidized form of the protein releasing the proton to the other side. Hence, the reduced form of the protein must be in a conformation that is distinct from the oxidized form (B). All electrons have one negative charge and therefore have the same affinity for protons (A). More than one proton can be pumped per electron that passes through a complex (C), depending on how many different carriers are in the complex and how much energy is released by the transfer of electrons through the entire complex. Since protons are readily available from any aqueous solution, the reducing agent does not need to donate protons to the transport complex (D).

A13–15.   C. By definition, E_0' refers to the standard state of equal concentrations of each member of the redox pair. Therefore $\Delta E_0'$ does not vary with the actual concentrations. Compounds with positive redox potentials can donate electrons to other compounds under standard conditions, so long as the electron acceptor has a higher redox potential. Compounds that are able to donate only one electron do not necessarily have higher redox potentials than those of compounds that are able to donate two electrons. (Water, for example, has a very high redox potential.) Although the $\Delta E_0'$ of a reaction is directly proportional to the $\Delta G^{\circ\prime}$ of a reaction and both are independent of the concentrations of substrates and products, the ΔG depends on these concentrations. In option E, we cannot tell from the E_0 values whether or not the gain of electrons by water from AH_2 is favorable, since the value of 820 mV refers only to the *loss* of electrons by water.

A13–16. B and E. In order for a reaction to drive ATP synthesis under standard conditions, the $\Delta G°'$ of the reaction must be less than –7.3 kcal/mol. Since $\Delta G°' = -n\ (0.023)\ \Delta E'_0$, the value of $\Delta E'_0$ must be greater than 317 mV/n, where n is the number of electrons transferred. $\Delta E'_0$ is 130 mV for the reduction of a molecule of pyruvate by NADH, 390 mV for the reduction of a molecule of cytochrome b by NADH, 40 mV for the reduction of a molecule of cytochrome b by ubiquinone, 200 mV for the oxidation of a molecule of ubiquinone by cytochrome c, and 590 mV for the oxidation of cytochrome c by oxygen. The number of electrons transferred in each of the above reactions is 2, 1, 1, 1, and 1, respectively. Thus only reactions B and E are sufficient to drive ATP synthesis.

A13–17. E. Cytochrome oxidase, which is the last carrier in the mitochondrial electron-transport chain and thus has the highest redox potential, contains copper ions and a heme group. Ubiquinone is not a protein and does not contain a metal group (A). Both 2Fe2S and 4Fe4S centers carry one electron (B). Iron-sulfur centers generally have a lower redox potential than do cytochromes (C). The heme group in cytochrome c contains a charged iron ion. The interiors of proteins are often hydrophobic, favoring a relatively high redox potential, since reduction of the iron ion decreases its charge, and charges are energetically unfavorable in a hydrophobic environment. Mutation of hydrophobic residues near the heme group of cytochrome c to acidic (negatively charged) amino acids will neutralize the charge on the iron atom and most likely decrease the redox potential, by making the uptake of an electron less energetically favorable (D).

A13–18. Oxygen is held tightly by the cytochrome oxidase of the respiratory chain until it has accepted the four electrons (plus the 4 H^+) required to convert it to 2 molecules of water. Superoxide is therefore never released.

A13–19. A, B, and E. The thylakoid spaces are continuous with one another, and grana are merely stacks of thylakoids, so ions can flow from one thylakoid space to another and from one granum to another. Since the outer membrane is porous, ions are also able to freely diffuse from the cytosol to the intermembrane space.

A13–20. Light reactions: $H_2O + NADP^+ + H^+ + ADP + P_i \rightarrow O_2 + NADPH + ATP$. Dark reactions: $CO_2 + NADPH + ATP \rightarrow (CH_2O) + NADP^+ + H^+ + ADP + P_i$. (CH_2O) indicates a one-carbon unit of a carbohydrate.

A13–21. B. The rate of photosynthesis will increase with increasing light intensity until all of the reaction centers are being hit directly by photons. At saturating levels of light, the rate of photosynthesis is limited by the number of reaction centers that are still capable of being excited and hence can be increased only by increasing the number of reaction centers or by increasing the rate at which the reaction centers are restored to their low energy state. Increasing the number of chlorophyll molecules in the antennae complexes, the energy per photon of light, or the rate at which chlorophyll molecules are able to transfer energy electrons to one another, will have no effect on either of these parameters. Adding a powerful oxidizing agent might, if anything, interfere with the reduction of the reaction center back to its resting state.

A13–22. C. If you now inhibit photosystem II, you will deprive plastoquinone, which can still donate its electrons to the b_6-f complex, of an electron source. Hence, plastoquinone will accumulate in its oxidized form. In contrast, all of the other components downstream of plastoquinone will be able to cycle between their oxidized and reduced states, although no net reduction can occur since there is no ultimate electron acceptor. ATP synthesis will continue, since electrons are still being fed through the b_6-f complex, and in fact the same amount of ATP will be generated, only now the photons used to drive ATP synthesis are those absorbed by photosystem I, rather than those absorbed by photosystem II.

A13–23.   A. In the carbon fixation process in chloroplasts, carbon dioxide is initially added to the sugar ribulose 1,5-bisphosphate.

B. The final product of carbon fixation in chloroplasts is the three-carbon compound glyceraldehyde 3-phosphate.

C. This is converted into pyruvate, which can be directly used by the mitochondria, into sucrose, which is exported to other cells, and into starch, which is stored in the stroma.

D. The carbon fixation cycle requires energy in the form of ATP and reducing power in the form of NADPH.

A13–24.   E. Three molecules of O_2 are required to form 3 molecules of 3-phosphoglycerate and 3 molecules of phosphoglycolate. In order to break even (i.e., simply to keep the Calvin cycle going), you need to have enough 3-phosphoglycerate to synthesize ribulose 1,5-bisphosphate again. Therefore, for every 3 molecules of O_2 that react with ribulose 1,5-bisphosphate, you need to generate 2 additional molecules of 3-phosphoglycerate. For every 3 molecules of CO_2 that go into the Calvin cycle, one molecule of 3-phosphoglycerate is formed. So if you have at least 6 molecules of CO_2 per 3 molecules of O_2 going through the Calvin cycle, you will break even. Only if you have a ratio of CO_2 to O_2 higher than 6:3 (2:1) can you have a net synthesis of carbohydrate.

A13–25.   C. Mitochondria are most closely related to *Bacillus*, and chloroplasts to cyanobacteria. Maize (a eucaryote) is more closely related to *Giardia* (a simple eucaryote) than it is to bacteria (procaryotes).

A13–26.   A and C. *Methanococcus jannaschii* is an archaebacterium that lives in very hot, anaerobic thermal vents that are thought to resemble the conditions under which the first cells lived. Like many organisms that live under these unpleasant conditions, *Methanococcus jannaschii* seems to be more closely phylogenetically related to the proposed ancestor cell than it is to most other organisms. However, the genome of *Methanococcus* is perhaps 18 times larger than the genome of the suspected ancestor cell, which may have had only 100 or so genes at most. The reactions that *Methanococcus* uses in carbon fixation and energy generation use many proteins that are different from those used by chloroplasts, cyanobacteria, mitochondria, and aerobic bacteria.

14 Intracellular Compartments and Transport

Questions

MEMBRANE-BOUNDED ORGANELLES (Pages 448–452)
Eucaryotic Cells Contain a Basic Set of Membrane-bounded Organelles (Pages 449–450)

14–1   Easy, short answer

Name the membrane-bounded compartments in a eucaryotic cell where each of the functions listed below takes place.

    A. Photosynthesis.

    B. Transcription.

    C. Oxidative phosphorylation.

    D. Modification of secreted proteins.

    E. Steroid hormone synthesis.

    F. Degradation of worn-out organelles.

    G. New membrane synthesis.

    H. Breakdown of lipids and toxic molecules.

14–2   Easy, multiple choice

Which organelles are most numerous in a typical animal cell such as a liver cell?

    A. Peroxisomes.

    B. Lysosomes.

    C. Endosomes.

    D. Mitochondria.

    E. Golgi apparatus.

14–3   Easy, multiple choice

Which membrane-bounded compartment occupies the greatest volume (in total) in a typical animal cell such as a liver cell?

    A. Nucleus.

    B. Cytosol.

    C. Mitochondria.

    D. Lysosomes.

    E. Endoplasmic reticulum.

Membrane-bounded Organelles Evolved in Different Ways (Pages 450–452)

14–4   Easy, short answer

You discover a fungus that contains a strange star-shaped organelle not found in any other eucaryotic cell you have seen. On further investigation you find the following:

    A. the organelle possesses a small genome in its interior.

    B. the organelle is surrounded by two membranes.

    C. vesicles do not pinch off the organelle membrane.

    D. the interior of the organelle contains proteins similar to those of many bacteria.

    E. the interior of the organelle contains ribosomes.

How might this organelle have arisen?

14–5   Easy, art labeling

Complete diagrams B and C in Figure Q14–5 to show how the nucleus and endoplasmic reticulum might have evolved in an ancient procaryotic ancestor of the eucaryotic cell. (You do not need to sketch any of the other organelles in the cell.) Label the different features (membranes, compartments) in the ancient eucaryotic cell.

Q14–5

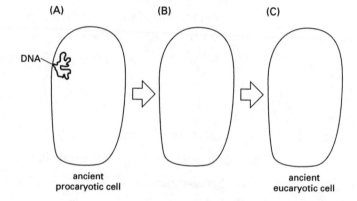

PROTEIN SORTING (Pages 452–462)

Proteins Are Imported into Organelles by Three Mechanisms (Page 453)

14–6   Easy, matching/fill in blanks

For each of the following sentences, fill in the blanks with the correct word or phrase selected from the list below. Use each word or phrase only once.

    A. Plasma membrane proteins are inserted into the membrane in the _____.

    B. The address information for protein sorting in a eucaryotic cell is contained in the _____ of the proteins.

    C. Proteins enter the nucleus in their _____ form.

    D. Proteins that remain in the cytosol do not contain a _____.

    E. Proteins are transported into the Golgi apparatus via _____.

    F. The proteins transported into the endoplasmic reticulum by _____ are in their _____ form.

protein translocators; amino acid sequence; endoplasmic reticulum; plasma membrane; sorting signal; folded; unfolded; transport vesicles; Golgi apparatus.

Signal Sequences Direct Proteins to the Correct Compartment (Pages 453–455)

14–7 Intermediate, short answer (Requires information from section on pages 458–459)

What would happen in each of the following cases? Assume in each case that the protein involved is a soluble protein, not a membrane protein.

(A) You add a signal sequence (for the ER) to the amino-terminal end of a normally cytosolic protein.

(B) You change the hydrophobic amino acids in an ER signal sequence into charged amino acids.

(C) You change the hydrophobic amino acids in an ER signal sequence into other, hydrophobic, amino acids.

(D) You move the amino-terminal ER signal sequence to the carboxyl-terminal end of the protein.

14–8 Intermediate/difficult, short answer

You are trying to identify the peroxisome-targeting sequence in the peroxisomal enzyme thiolase from yeast. To this end, you have created a variety of hybrid genes that encode hybrid proteins containing part of the thiolase attached to another protein, histidinol dehydrogenase (HDH), which is a cytosolic enzyme required for the synthesis of the amino acid histidine. You genetically engineer a series of yeast cells to express these hybrid proteins instead of their own versions of the enzymes. If the hybrid protein is imported into the peroxisome, the HDH portion cannot function properly and the yeast is unable to grow on medium lacking histidine. You obtain the results shown in Figure Q14–8. Where are the peroxisomal targeting sequence(s) in thiolase?

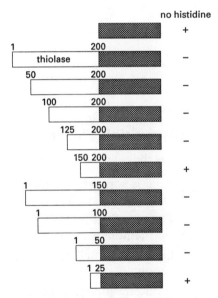

The numbers above each protein indicate the amino acids from thiolase present in the hybrid protein.
+ indicates growth on medium lacking histidine;
− indicates lack of growth on medium lacking histidine.

Q14–8

Proteins Enter the Nucleus Through Nuclear Pores (Pages 455–457)

14–9 Easy, multiple choice

Which of the following are routinely both imported into and exported from the nucleus?

A. Histones.

B. Gene regulatory proteins.

C. DNA polymerases.

D. Ribosomal proteins.

E. Nuclear lamina proteins.

14–10   Intermediate, multiple choice

The cytoplasms of adjoining plant cells are connected by fine channels called plasmodesmata whose structure is shown in Figure Q14–10:

Movement of proteins through plasmodesmata is likely to be most similar to movement of proteins:

    A. from the cytoplasm into a mitochondrion.

    B. from the cytoplasm into the ER.

    C. from the ER to the Golgi aparatus.

    D. from the nucleus into the cytoplasm.

    E. from the cytoplasm into a peroxisome.

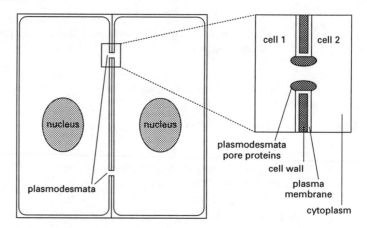

Q14–10

14–11   Easy, multiple choice

What is the role of the nuclear localization sequence in a nuclear protein?

    A. It is bound by cytoplasmic proteins that direct the nuclear protein to the nuclear pore.

    B. It is a hydrophobic sequence that enables the protein to enter the nuclear membranes.

    C. It aids protein unfolding in order for the protein to thread through nuclear pores.

    D. It prevents the protein diffusing out of the nucleus via nuclear pores.

    E. It directs the protein to the nuclear lamina.

14–12   Difficult, data interpretation (Requires information from Chapter 5 panels)

A gene regulatory protein, A, contains a typical nuclear localization signal but suprisingly is usually found in the cytosol of a cell. When the cell is exposed to hormones, however, protein A moves from the cytoplasm into the nucleus where it activates genes involved in cell proliferation. When you purify protein A from unstimulated cells, you find that another protein, protein B, is complexed with it. To determine the function of protein B, you make mutants lacking the gene for protein B. You then fractionate extracts of gently lysed normal and B mutant cells into cytoplasmic and nuclear fractions, separate the proteins in these fractions by gel electrophoresis, and probe the gels for the presence of proteins A and B using Western blotting techniques. Your results are shown in Figure Q14–12.

On the basis of these results, which is the most likely function of Protein B? Explain your reasoning.

   A. In the presence of hormone, Protein B neutralizes the charges on Protein A's nuclear localization signal by covalently attaching hydrophobic groups to the lysine side chains.

   B. In the absence of hormone, protein B cleaves the nuclear localization signal off protein A.

   C. In the absence of hormone, Protein B binds to the nuclear localization signal on protein A, blocking its action.

   D. Protein B is a nuclear import receptor.

   E. Protein B prevents Protein A from unfolding.

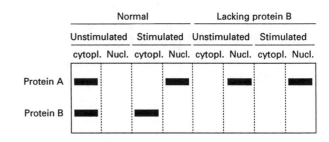

Q14–12

Proteins Unfold to Enter Mitochondria and Chloroplasts (Pages 457–458)

14–13   Easy, multiple choice

Which of the following statements about import of proteins into mitochondria are true?

   A. The signal sequences on mitochondrial proteins are usually carboxyl terminal.

   B. The first stage of import of a mitochondrial protein is across the outer membrane into the intermembrane space.

   C. Most mitochondrial proteins are not imported from the cytosol but are synthesized inside the mitochondria.

   D. Mitochondrial proteins are translocated across the inner and outer membranes simultaneously.

   E. Mitochondrial proteins cross the membrane in an unfolded state.

14–14   Intermediate/difficult, multiple choice (Requires information from Chapter 5)

Which of the following will inhibit import of an enzyme into mitochondria?

   A. Chaperone proteins.

   B. A small molecule that binds tightly to the active site of the enzyme.

   C. Inhibition of ATP synthesis.

   D. A high concentration of free mitochondrial signal peptide in the cytosol.

   E. An inhibitor of mitochondrial signal peptidase.

Proteins Enter the Endoplasmic Reticulum While Being Synthesized (Pages 458–459)

14–15   Easy, multiple choice

Proteins destined to enter the endoplasmic reticulum:

A. are transported across the membrane after their synthesis is complete.

B. are synthesized on free ribosomes in the cytosol.

C. begin to cross the membrane while still being synthesized.

D. cross the membrane in a folded state.

E. all remain within the endoplasmic reticulum.

14–16   Easy, multiple choice

After isolating the rough endoplasmic reticulum from the rest of the cytoplasm, you purify the RNAs attached to it. What proteins do you expect these RNAs to encode?

A. Soluble secreted proteins.

B. ER membrane proteins.

C. Mitochondrial membrane proteins.

D. Plasma membrane proteins.

E. Ribosomal proteins.

Soluble Proteins Are Released into the ER Lumen (Pages 459–460)

14–17   Intermediate, multiple choice + short answer

You are studying the *in vitro* translation and import of proteins into isolated vesicles made from the ER (microsomes). Using differential centrifugation you have separated the cytoplasm of the cell type from which the ER originally came into several different fractions. One of these fractions stimulates the import of proteins into microsomes if the microsomes are added after translation is completed. It has no effect on protein import when translation takes place in the presence of the microsomes. Which of the following are most likely to be present in this fraction? Explain your answer.

A. Nuclear import receptor.

B. SRP receptor.

C. Signal peptidase.

D. Chaperone protein.

E. Ribosomes.

Start and Stop Signals Determine the Arrangement of a Transmembrane Protein in the Lipid Bilayer (Pages 461–462)

14–18   Easy, short answer

Briefly describe the mechanism by which the presence of an internal stop-transfer sequence in a protein causes the protein to become embedded in the lipid bilayer as a transmembrane protein with a single-membrane-spanning region? Assume that the protein has an amino terminal signal sequence and just one internal hydrophobic stop-transfer sequence.

14–19 Intermediate/difficult, art labeling

A plasma membrane protein X traverses the membrane three times in the orientation shown in Figure Q14–19. (N = amino terminus; C = carboxyl terminus.) The hydrophobic membrane-spanning regions are shown as open boxes.

(A) Sketch the arrangement of the newly synthesized protein chain after it has completed its entry into the ER membrane but before any action of signal peptidase. Label the cytosol and ER lumen in your diagram.

(B) Label the location of stop-transfer (S) and start-transfer (T) sequences on your diagram.

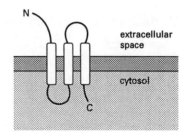

Q14–19

14–20 Intermediate/difficult, multiple choice (Note to instructors: this is an alternative question to 14–19 that does not require the student to sketch anything; giving them both in the same test will give away the answer to 14–19)

Figure Q14–20B shows the orientation of a multipass membrane protein in the plasma membrane. This protein also had an amino-terminal signal sequence, which was cleaved off in the endoplasmic reticulum (Figure Q14–20A). All of the membrane-spanning regions (depicted as open boxes) in this protein have the same amino acid sequence, leading to the hypothesis that any hydrophobic signal sequence has the potential to act as either a start or a stop signal; the actual behavior of the signal is determined by the position of the signal relative to other signals in the protein. If this hypothesis were true, which of the following modifications would cause the protein depicted in Figure Q14–20B to be inserted in the membrane in exactly the opposite orientation relative to the cytoplasm?

A. Changing hydrophobic amino acids in the first, cleaved, sequence to charged amino acids.

B. Changing hydrophobic residues in every other signal sequence to charged residues, starting with the first, cleaved, signal sequence.

C. Adding a new signal sequence and signal peptidase recognition site to the carboxyl terminus of the protein.

D. Deleting the first signal sequence.

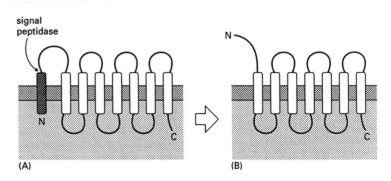

Q14–20   (A)   (B)

VESICULAR TRANSPORT (Pages 462–467)
Transport Vesicles Carry Soluble Proteins and Membrane Between Compartments (Page 463)

14–21   Easy, matching/fill in blanks

For each of the following sentences, fill in the blanks with the correct word or phrase selected from the list below. Use each word or phrase only once.

A. Proteins are transported out of a cell via the _____ or _____ pathway.

B. Fluid and macromolecular material is transported into the cell via the _____ pathway.

C. All proteins being transported out of the cell pass through _____ and _____.

D. Transport vesicles link organelles of the _____ system.

E. The formation of _____ in the endoplasmic reticulum stabilizes protein structure.

the endoplasmic reticulum; endomembrane; carbohydrate; secretory; endocytic; the Golgi apparatus; endosome; lysosome; protein; disulfide bonds; exocytic; hydrogen bonds; ionic bonds.

14–22   Easy, multiple choice

An individual transport vesicle:

A. contains only one type of protein in its lumen.

B. will fuse with only one type of membrane.

C. is endocytic if it is traveling toward the plasma membrane.

D. is enclosed by a membrane with the same lipid and protein composition as the membrane of the donor organelle.

Vesicle Budding Is Driven by the Assembly of a Protein Coat (Pages 463–465)

14–23   Easy, multiple choice

Clathrin-coated vesicles do NOT:

A. bud from the plasma membrane.

B. bud from the endoplasmic reticulum.

C. bud from the Golgi apparatus.

D. require GTP hydrolysis for their budding.

E. contain adaptin coat proteins.

14–24   Easy, multiple choice

Adaptin proteins:

A. are found in both clathrin-coated and COP vesicles.

B. can hydrolyze GTP.

C. form a basketlike network on the cytosolic side of the membrane.

D. remain on the vesicle after clathrin coat disassembly.

E. bind to different types of cargo receptors.

14–25   Intermediate, short answer

Clathrin-coated vesicles will bud from plasma membrane fragments when adaptins, clathrin, and dynamin-GTP are added. What would you observe in each case if you omitted:

(A) dynamin?

(B) adaptins?

(C) clathrin?

The Specificity of Vesicle Docking Depends on SNAREs (Pages 465–467)

14–26   Easy, multiple choice

vSNARES participate directly in:

A. assembly or formation of the transport vesicles.

B. movement of the vesicle along cytoskeletal filaments.

C. uncoating of the vesicle.

D. docking of the vesicle to the target organelle.

E. fusion of the vesicle to the target organelle.

SECRETORY PATHWAYS (Pages 467–472)
Most Proteins Are Covalently Modified in the ER (Pages 467–468)

14–27   Easy, multiple choice

N-linked oligosaccharides on secreted glycoproteins are attached to:

A. nitrogen atoms in the polypeptide backbone.

B. the serine or threonine in the sequence Asn-X-Ser/Thr.

C. the amino terminus of the protein.

D. the asparagine in the sequence Asn-X-Ser/Thr.

E. the aspartic acid in the sequence Asp-X-Ser/Thr.

14–28   Easy, multiple choice

Which of the following statements about glycosylation are true?

A. Cytosolic proteins are never glycosylated.

B. Sugars are transferred one by one from dolichol to protein.

C. The oligosaccharide transferred from dolichol is composed only of mannose.

D. The oligosaccharide on a protein leaving the endoplasmic reticulum usually differs from the oligosaccharide transferred to that protein from dolichol.

E. Glycosylation occurs only after the protein has completed its entry into the endoplasmic reticulum.

Proteins Are Further Modified and Sorted in the Golgi Apparatus (Pages 469–470)

14–29   Easy, art labeling

Match the set of labels below with the numbered label lines on Figure Q14–29.

A. Cisterna.

B. Golgi stack.

C. Secretory vesicle.

D. *trans* Golgi network.

E. *cis* Golgi network.

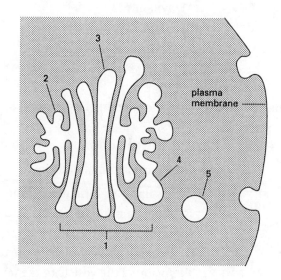

Q14–29

14–30   Intermediate/difficult, data interpretation

A plasma membrane protein carries an oligosaccharide containing mannose (Man), galactose (Gal), sialic acid (SA), and *N*-acetylglucosamine (GlcNAc). These sugars are added to the protein as it proceeds through the secretory pathway. First, a core oligosaccharide containing Man and GlcNAc is added, followed by Gal, Man, SA, and GlcNAc in a particular order. Each addition is catalyzed by a different transferase acting at a different stage as the protein proceeds through the secretory pathway. You have isolated mutants defective for each of the transferases, purified the membrane protein from each of the mutants, and identified which sugars are present in each mutant protein. The results are summarized in Table Q14–30.

Table Q14–30

Cell lacking:	Sugars present in the purified protein			
	Man	Gal	SA	GlcNAc
A. Oligosaccharide protein transferase	–	–	–	–
B. Galactose transferase	+	–	–	+
C. SA transferase	+	+	–	+
D. GlcNAc transferase	+	–	–	less than in normal cells

From these results, match each of the transferases (A, B, C, D) to its subcellular location select-ed from the list below. (Assume that each location contains only one enzyme.)

1. Central Golgi cisternae.
2. *cis* Golgi network.
3. ER.
4. *trans* Golgi network.

14–31   Easy, matching/fill in blanks

Complete each of the following sentences correctly by crossing out one of the alternatives in brackets.

A. New plasma membrane reaches the plasma membrane by the [regulated / constitutive] exocytosis pathway.
B. New plasma membrane proteins reach the plasma membrane by the [regulated / constitutive] exocytosis pathway.
C. Insulin is secreted from pancreatic cells by the [regulated / constitutive] exocy-tosis pathway.
D. The interior of the *trans* Golgi network is [acidic / alkaline].
E. Proteins that are constitutively secreted [aggregate / do not aggregate] in the *trans* Golgi network.

Secretory Proteins Are Released from the Cell by Exocytosis (Pages 470–472)

14–32   Intermediate, multiple choice

You have purified insulin-containing vesicles from pancreatic cells and exocytic vesicles from another cell type not specialized for secretion. In which of the following ways do you expect the two types of vesicles to be similar?

A. The same tSNAREs mark the cytoplasmic face of the vesicle membranes.
B. They will have the same concentration of protein in their lumens.
C. They are likely to have a similar concentration of Ca^{2+} in their lumens.
D. The same cytoplasmic fractions must be added in order to stimulate fusion with the plasma membrane.
E. They will be the same size.

ENDOCYTIC PATHWAYS (Pages 472–477)
Specialized Phagocytic Cells Ingest Large Particles (Pages 472–473)
Fluid and Macromolecules Are Taken Up by Pinocytosis (Pages 473–474)

14–33   Easy, multiple choice

Which of the following statements are true?

A. Pinocytosis and phagocytosis are the two major forms of exocytosis.
B. All eucaryotic cells ingest fluid by pinocytosis and do so at approximately the same rate.

C. Macrophages will endocytose large particles only if the particles bind to receptors on the phagocytic cell surface.

D. Both phagocytic and pinocytic vesicles begin assembly as clathrin-coated pits.

E. Phagocytic vesicles fuse with the endosome; pinocytic vesicles with the Golgi apparatus.

Receptor-mediated Endocytosis Provides a Specific Route into Animal Cells (Pages 474–475)

14–34 Difficult, data interpretation + multiple choice (Requires information from section on pages 475–476)

You have always had trouble keeping your blood cholesterol levels within a healthy range, whereas your sister has always been able to eat what she liked with impunity. To find out the reason, you compare the ability of your cells to take up LDL compared to your sister's cells. To do this you add LDL particles from your sister's blood to a sample of your sister's cells and to a sample of your own cells, and then wash off the LDL that has not become bound to the LDL receptor. You then wait either 2 minutes or 40 minutes, and treat half the cells in each sample with a protein-digesting enzyme (a protease). You then lyse the treated cells and determine the presence or absence of LDL protein. The results are given in Table Q14–34.

Table Q14–34

	After 2 minutes		After 40 minutes	
	Sister	You	Sister	You
Treated with protease	+	+	–	–
Not treated with protease	+	+	–	+

+ indicates LDL protein detected; – indicates LDL protein not detected

On the basis of these results, which of the following is the most likely explanation for your tendency to a high blood cholesterol? Explain your answer.

A. Although you did not test binding of your own LDL particle, you can deduce the only defect you could have must be a mutation in your own LDL protein that prevents it from binding to the receptor.

B. Your cells have fewer LDL receptors than do your sister's cells.

C. Your LDL receptors cannot bind to LDL.

D. Your LDL receptors bind to LDL but cannot be localized to clathrin-coated pits and internalized.

E. Your LDL receptors cannot release LDL at low pH.

Endocytosed Macromolecules Are Sorted in Endosomes (Pages 475–476)

14–35 Easy, short answer

Name three possible fates for an endocytosed molecule that has reached the endosome.

Lysosomes Are the Principal Sites of Intracellular Digestion (Pages 476–477)

14–36   Difficult, data interpretation

Fibroblast cells from patients W, X, Y and Z, who each have a different inherited defect, all contain "inclusion bodies," which are lysosomes filled with undigested material. You wish to identify the cellular basis of these defects. The possibilities are:

1. a defect in one of the lysosomal hydrolases.
2. a defect in the phosphotransferase that is required for mannose-6-phosphate tagging.
3. a defect in the mannose-6-phosphate receptor, which binds lysosomal proteins in the *trans* Golgi network and delivers them to lysosomes.

You find that when some of these mutant fibroblasts are incubated in media in which normal cells have been grown, the inclusion bodies disappear. This leads you to suspect that lysosomal hydrolases can be secreted by the constitutive exocytic pathway in normal cells and are being taken up by the mutant cells. (It is known that some mannose-6-phosphate receptor molecules are found in the plasma membrane and can take up and deliver lysosomal proteins via the endocytic pathway.) You incubate cells from each patient with media from normal cells and media from each of the other mutant cell cultures, and get the following results.

	Media				
Cell Line	From normal cells	From cultures of W cells	From cultures of X cells	From cultures of Y cells	From cultures of Z cells
Normal	+	+	+	+	+
W	–	–	–	–	–
X	+	+	–	–	–
Y	+	+	–	–	+
Z	+	+	–	+	–

+ indicates that the cells appear normal; – indicates that the cells still have inclusion bodies.

For each patient (W, X, Y, Z) indicate which of the defects (1, 2, 3) they are most likely to have.

Answers

A14–1.   A. Photosynthesis = chloroplast.

B. Transcription = nucleus.

C. Oxidative phosphorylation = mitochondrion.

D. Modification of secreted proteins = Golgi apparatus and rough endoplasmic reticulum (ER).

E. Steroid hormone synthesis = smooth ER.

F. Degradation of worn-out organelles = lysosome.

G. New membrane synthesis = ER.

F. Breakdown of lipids and toxic molecules = peroxisome.

A14–2.   D.

A14–3.   B.

A14–4.   A genome, a double membrane, ribosomes, and proteins similar to those found in bacteria are evidence for an organelle having evolved from an engulfed bacterium.

A14–5.   Figure A14–5.

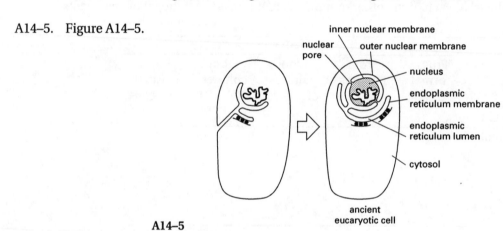

A14–5

A14–6.   A. Plasma membrane proteins are inserted into the membrane in the <u>endoplasmic reticulum</u>.

B. The address information for protein sorting in a eucaryotic cell is contained in the <u>amino acid sequence</u> of the proteins.

C. Proteins enter the nucleus in their <u>folded</u> form.

D. Proteins that remain in the cytosol do not contain a <u>sorting signal</u>.

E. Proteins are transported into the Golgi apparatus via <u>transport vesicles</u>.

F. The proteins transported into the endoplasmic reticulum by <u>protein translocators</u> are in an <u>unfolded</u> form.

A14–7.   (A) The protein will now be transported into the ER lumen. (B) The altered signal sequence will not be recognized and the protein will remain in the cytosol. (C) The protein will still be delivered into the ER. It is the distribution of hydrophobic amino acids that is important, not the actual sequence. (D) The protein will not enter the ER. Because the carboxyl terminus of the protein is the last part to be made, the ribosomes synthesizing this protein will not be recognized by the SRP and carried to the ER.

A14–8.   From the results, the sequences 1–50 and 125–200 are the smallest pieces of thiolase that are sufficient to cause HDH to be transported into the peroxisome; hence each of these pieces must contain a signal sequence. It is important to realize that even though 1–25 and 150–200 are not sufficient on their own to cause import into the peroxisome, we cannot tell from these data whether these sequences are necessary for import as part of the larger sequence. In other words, we cannot say that 25–50 and 125–150 contain complete signal sequences as a signal sequence might be located, for example, from 20–30 or from 145–155. Thus it is not possible to identify the target minimum sequence(s) from this experiment.

A14–9.   D. Ribosomal proteins are imported from the cytoplasm, where they are synthesized, and exported as part of ribosomal subunits, which are assembled in the nucleus.

A14–10.   D. Transport across the plasmodesmata involves crossing two contiguous membranes through a protein complex large enough to see in the electron microscope. This is most similar to transport through nuclear pores.

A14–11.   A.

A14–12.   C. Protein B could work by preventing the nuclear localization signal from binding to nuclear import receptors; this could be accomplished by either reversibly modifying the signal or by binding to the signal and shielding it from the nuclear import receptor. Since protein A is stably associated with protein B in unstimulated cells, C is the most likely answer. In unstimulated cells, both protein A and protein B are associated with one another in the cytoplasm. In stimulated cells, protein A but not protein B moves into the nucleus. In cells lacking protein B, protein A is nuclear both in the presence and absence of hormone. This tells us that protein B is required for keeping protein A out of the nucleus. Hence, it seems unlikely that protein B is a nuclear import receptor. Protein B is unlikely to act by preventing protein A from unfolding because nuclear import does not require unfolding. If protein B acted by cleaving the nuclear localization signal off protein A, we would expect cytoplasmic protein A to be smaller than nuclear protein A and thus show up as a band in a slightly different position in the gel. In addition, it would be extremely unlikely that protein A could regain a nuclear localization signal upon hormone stimulation.

A14–13.   D and E.

A14–14.   B, C, and D. Proteins must unfold in order to cross the mitochondrial membrane. Since the enzyme active site is a property of the folded protein, a small molecule that binds tightly to the active site (B) will stabilize the folded protein, prevent unfolding and inhibit import. Mitochondrial import is an active process and therefore requires ATP (C). A large concentration of the signal peptide would compete with the full-length enzyme for the receptor and should therefore inhibit import (D). Chaperone proteins (A) promote both folding and unfolding of the polypeptide chain and therefore would be expected to promote import of the enzyme. Cleavage of the signal is not required for import, so inhibition of the signal peptidase will not affect import (E).

A14–15.   C.

A14–16.   A, B, and D. The rough ER consists of ER membranes and polyribosomes that are in the process of translating and translocating proteins into the ER membrane and lumen. Thus all proteins that end up in the lysosome, Golgi apparatus, or plasma membrane, or are secreted, will be encoded by the RNAs associated with the rough ER. Mitochondrial and ribosomal proteins are translated on free cytosolic ribosomes.

A14–17.   D. When the microsomes are added to proteins that have already been translated, the proteins will have already folded and would need to unfold before they could be imported into the ER. Hence, chaperones might be expected to facilitate import. They are not required when microsomes are present during translation since the proteins that are being translated are immediately imported as they come off the ribosome before protein folding can take place. The presence of nuclear import receptors (A) is irrelevant for protein import into the ER. SRP receptor (B) is a membrane protein and is present in the microsomes. Signal peptidase (C) is not required for import into the lumen of the ER; it is required only for the final processing of the protein. Ribosomes (E) will have no effect on proteins that have already been translated.

A14–18.   The amino-terminal signal sequence initiates translocation and the protein chain starts to thread through the translocation channel. When the stop-transfer sequence enters the translocation channel, the channel discharges both the signal sequence and the stop-transfer sequence sideways into the lipid bilayer. The signal sequence is then cleaved, so that the protein remains held in the membrane by the hydrophobic stop-transfer sequence.

A14–19.   (A) Figure A14–19. Since the amino terminus (N) of the protein is on the extracellular side of the plasma membrane, there must have been a signal sequence (acting as a start signal) at the extreme amino terminus; this signal was cleaved off by signal peptidase in the ER lumen. (B) Figure A14–19.

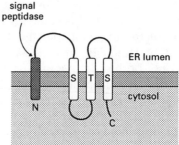

A14–19

A14–20.   D. Figure A14–20A. Deleting the first signal sequence completely would convert the original first stop-transfer signal to an internal start-transfer signal and would both invert the protein in the membrane and get rid of the signal sequence that is not present in the original fully processed protein. Changing hydrophobic amino acids to charged amino acids both destroys the ability of the sequence to act as a signal sequence and to become a membrane-spanning sequence. So, mutating the first signal sequence so that it loses its function (A) will also cause the second signal sequence to become an internal start-transfer sequence. In this case, however, the extreme amino terminus would remain on the cytoplasmic side of the membrane and the mutated signal sequence would not be cleaved off because signal peptidase is found only inside the ER (Figure A14–20B). Mutating every other signal sequence (B) so that it loses its function would decrease the number of transmembrane regions and increase the size of the internal loops between membrane-spanning regions (Figure A14–20C). Adding a new signal sequence to the carboxyl terminus of the protein (C) will have no effect on the orientation of the protein in the membrane, since the amino terminus will have been inserted before the final signal is even translated.

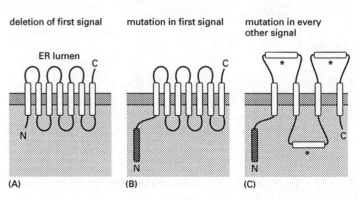

A14–20

A14–21.   A. Proteins are transported out of a cell via the <u>secretory</u> or <u>exocytic</u> pathway.

B. Fluid and macromolecules are transported into the cell via the <u>endocytic</u> pathway.

C. All proteins being transported out of the cell pass through <u>the endoplasmic reticulum</u> and <u>the Golgi apparatus</u>.

D. Transport vesicles link organelles of the <u>endomembrane</u> system.

E. The formation of <u>disulfide bonds</u> in the endoplasmic reticulum stabilizes protein structure.

A14–22.   B. An individual vesicle may contain more than one type of protein in its lumen (A), all of which will contain the same sorting signal (or will lack specific sorting signals). Endocytic vesicles (C) generally move away from the plasma membrane. The vesicle will not necessarily contain the same lipid and protein composition as the donor organelle, since the vesicle is formed from a selected subset of organelle membrane and contents from which it budded (D).

A14–23.   B. COP proteins are thought to coat the vesicles used in ER to Golgi transport. Clathrin coats form around vesicles budding from the Golgi and from the plasma membrane (A and C). GTP hydrolysis is required for dynamin-induced pinching off of the vesicle bud (D). They do also contain adaptin coat proteins (E).

A14–24.   E. Adaptins are found only in clathrin-coated vesicles (A). They cannot hydrolyze GTP (B); dynamin hydrolyzes GTP in clathrin-coated vesicles. Clathrin forms a basketlike network (C). They are shed with the clathrin coat (D).

A14–25.   (A) Clathrin-coated pits would form, but would not bud off. (B) Clathrin coats would not form around the vesicle. Adaptins are required to link the clathrin to the membrane. (C) Without clathrin, adaptins could still bind to the membrane, but no coat would form.

A14–26.   D. vSNAREs are proteins on the cytoplasmic face of transport vesicles that are used in target membrane recognition. vSNAREs are required for docking to the target membrane. Fusion of the two membranes is distinct from docking and is thought to be catalyzed by a separate group of proteins.

A14–27.   D.

A14–28.   D. Since the oligosaccharide chain is modified by enzymes in the ER and Golgi, most proteins leaving the Golgi do not contain the 14–sugar oligosaccharide that was originally transferred from dolichol to the protein. A few cytoplasmic proteins are glycosylated (A), albeit with a single sugar residue. Sugars are transferred as a single branched, preassembled oligosaccharide from dolichol to the protein (B). The oligossaccharide contains mannose, glucose, and *N*-acetylglucosamine (C). Glycosylation occurs as soon as the first suitable Asn enters the endoplasmic reticulum lumen (E).

A14–29.   1 = B; 2 = E; 3 = A; 4 = D; 5 = C.

A14–30.   A = 3 (oligosaccharide protein transferase = ER); B = 1 (galactose transferase = central Golgi cisternae); C = 4 (SA transferase = *trans* Golgi network); D = 2 (GlcNAc transferase = *cis* Golgi network). Proteins are modified in a stepwise fashion in the Golgi apparatus, with early steps taking place in the *cis* Golgi, intermediate steps taking place in the central Golgi cisternae, and late steps occurring in the *trans* Golgi network. If each enzyme produces the substrate for the next step, then a mutant lacking the enzyme that catalyzes the addition of the first sugar will be missing all of the sugars, a mutant lacking the enzyme that catalyzes the addition of the second sugar will contain the first sugar but will lack the other three, and so on. By this logic, mannose and GlcNAc must be the first sugars added, additional GlcNAc the second, galactose the third, and SA the last. Hence, the oligosaccharide protein transferase must be in the ER, the GlcNAc transferase in the *cis* Golgi, the galactose transferase in the central Golgi, and the SA transferase in the *trans* Golgi.

A14–31.   A. New plasma membrane reaches the plasma membrane by the <u>constitutive</u> exocytosis pathway.

B. New plasma membrane proteins reach the plasma membrane by the <u>constitutive</u> exocytosis pathway.

C. Insulin is secreted from pancreatic cells by the <u>regulated</u> exocytosis pathway.

D. The interior of the *trans* Golgi network is <u>acidic</u>.

E. Proteins that are constitutively secreted <u>do not aggregate</u> in the *trans* Golgi network.

A14–32.   C. Since ions do not have specific receptors, they are nonspecifically packaged into vesicles, so we expect both types of vesicle to have similar concentrations of Ca^{2+} as the donor organelle, the Golgi apparatus. tSNAREs mark the target membranes not the vesicles (A). The concentration of proteins in regulated vesicles is much higher due to the ability of the proteins packaged into such vesicles to aggregate (B). Regulated vesicles require additional stimulation in order to fuse with the plasma membrane *in vivo*, so we expect that these vesicles should also require the addition of an activating fraction or removal of an inhibitory fraction in order to fuse *in vitro* (D). Vesicles vary in size, but regulated secretory vesicles can be huge (relatively speaking), as one might expect, since they are formed around aggregates of protein (E).

A14–33.   C. Particles lacking surface molecules that are able to bind to macrophage receptors do not stimulate phagocytosis and are therefore not engulfed; this mechanism prevents macrophages from devouring the animal's own cells. Pinocytosis and phagocytosis are forms of endocytosis, not exocytosis (A). The rate of pinocytosis varies from cell type to cell type (B). Phagocytic vesicles start out as pseudopodia that are extended around the particle to be engulfed (D). Both pinocytic and phagocytic vesicles fuse with the endosome (E).

A14–34.   E. Protease treatment will digest any LDL protein bound to the cell surface. If the LDL protein is still present in the cell lysate after protease treatment, this indicates that it has been endocytosed, since the protease cannot cross the membrane. The results from the "2 minute" experiment show that both your cells and your sister's cells take up LDL identically. Therefore, you must have the same amount of LDL receptor as your sister (so B cannot be the case), and your receptors must be able to bind LDL, form clathrin-coated pits and be internalized (so C and D cannot be the case). The "40 minute" experiment shows that after this time, all the LDL taken up by your sister's cells has been degraded, as none is present, whether or not protease is added to the cell sample. This is most likely due to the LDL having been transferred to the lysosome and hydrolyzed. LDL can still be found in your cells, on the other hand. The results from your cells show that the LDL has not been degraded and instead has reappeared on the cell surface, since it is once again sensitive to the added protease. By permanently blocking receptors in this way it prevents efficient uptake of LDL from the blood; hence your tendency to high blood cholesterol. Although there is the possibility that you could also have a defect in your own LDL particles (not tested for), if that were your only defect (A), your cells and your sister's cells should behave identically at both time points.

A14–35.   1 recycled to the original membrane; 2 destroyed in the lysosome; 3 transcytosed across the cell to a different membrane.

A14–36.   W = 3 (defect in mannose-6-phosphate receptor); X = 2 (defect in phosphotransferase); Y = 1 and Z = 1 (defect in lysosomal hydrolases). These will be defects in two different lysosomal acid hydrolases. A cell that has no mannose-6-phosphate receptor will be able to make all the lysosomal hydrolases properly but will not be able to send them to the lysosome. Nor will it be able to scavenge hydrolases from the external media. Hence, this cell line cannot be rescued by culture media that has had lysosomal hydrolases secreted into it and thus will not be rescued by any of the media tested here. A cell line that has no phosphotransferase will be able to scavenge hydrolases from the external medium, but since all of the cell's own hydrolases will lack the mannose-6-phosphate tag, it will be rescued only by media from a cell line that is able to make all of the hydrolases. Cell lines missing one hydrolase will be rescued by media from any cell line that is able to secrete that hydrolase in a mannose-6-phosphate tagged form; in addition, media from cultures of cells missing a hydrolase will rescue any cell line with another type of defect.

15 Cell Communication

Questions

GENERAL PRINCIPLES OF CELL SIGNALING (Pages 482–493)
Signals Can Act over Long or Short Range (Pages 482–484)

15–1 Easy, matching/fill in blanks

Complete each of the following sentences by crossing out the incorrect words or phrases from the alternatives given in brackets.

A. Hormones are made by [endocrine / paracrine] cells and are carried to their target cells by [nerve cell processes / the bloodstream].

B. Paracrine signaling involves signal molecules known as [hormones / local mediators].

C. Neuronal signaling is a specialized example of [paracrine / contact-dependent] signaling.

D. If a cell cannot respond to a particular signal molecule, it may be because it does not carry a [receptor / gene] for that molecule.

E. In cell signaling, the conversion of the signal from one form into another is known as signal [transcription / translation / transduction].

15–2 Intermediate, multiple choice

In which of the following ways do hormones differ from paracrine signals?

A. Hormones are usually much less soluble in water than are paracrine signals.

B. Hormones are less stable than are paracrine signals.

C. Hormones have a lower affinity for their receptors than do paracrine signals.

D. Hormones are produced in greater quantities than are paracrine signals.

E. Hormones are produced by most cell types; paracrine signals are only produced in response to an insult such as an infection or a wound.

Each Cell Responds to a Limited Set of Signals (Pages 484–486)

15–3 Intermediate, data interpretation + multiple choice

Cell lines A and B both survive in tissue culture containing serum but do not proliferate. Addition of Factor F stimulates proliferation of cell line A but not cell line B. Cell line A produces a receptor protein (R), and, in order to see what the role of R might be, you introduce this receptor protein into cell line B, using recombinant DNA techniques. You then test all of your various cell lines in the presence of serum for their response to factor F, with the results summarized in Table Q15–3 (*next page*).

Table Q15–3

Cell line		Response
A	– Factor F	Cells do not proliferate
A	+ Factor F	Cells proliferate
B	– Factor F	Cells do not proliferate
B	+ Factor F	Cells do not proliferate
B + receptor R	– Factor F	Cells proliferate
B + receptor R	+ Factor F	Cells proliferate

Which of the following can be concluded from your results?

A. Binding of Factor F to its receptor is required for proliferation of cell line A but not cell line B.

B. Binding of receptor R to its ligand is required for proliferation of cell line B but not cell line A.

C. Receptor R probably does not bind to Factor F.

D. Cell line B does not normally have a receptor for Factor F.

Receptors Relay Signals via Intracellular Signaling Pathways (Pages 486–487)

15–4   Intermediate, multiple choice (Requires information from the whole chapter)

At which of the following steps in a signaling pathway can amplification of the signal occur?

A. An extracellular signal molecule binds and activates a cell-surface receptor.

B. An activated G protein activates adenylate cyclase, the enzyme that produces cyclic AMP.

C. Adenylate cyclase produces cyclic AMP.

D. Cyclic AMP activates a protein kinase (A-kinase).

E. A-kinase phosphorylates target proteins.

15–5   Intermediate, multiple choice

The major effect of hormone Q is to promote cell movement by increasing the activity of enzymes that are stimulated by high intracellular levels of Ca^{2+}. Upon binding to Q, the Q receptor interacts with the α subunit of a dimeric protein $\alpha\beta$, inducing a conformational change that frees the β subunit. The free β subunit then binds to and opens a Ca^{2+} channel in the plasma membrane. When Q dissociates from the receptor, the receptor releases α, which reassociates with β, thus causing the ion channel to close. In which of the following ways does this signaling pathway differ from a simpler mechanism in which Q binds directly to the ion channel?

A. The more complex pathway increases the speed with which the signal is converted to a response.

B. The more complex pathway more effectively transfers the signal to the place where the response takes place.

C. The more complex pathway amplifies the signal.

D. The more complex pathway allows the signal to simultaneously affect processes other than those affected by the concentration of Ca^{2+}.

E. The more complex pathway increases the opportunities for the response in the cytoplasm to be regulated by conditions both inside and outside the cell.

Some Signal Molecules Can Cross the Plasma Membrane (Pages 488–489)

15–6   Easy, multiple choice

All members of the steroid hormone receptor family:

A. are cell-surface receptors.

B. have enzymatic activity that is stimulated upon hormone binding.

C. are found only in the cytoplasm.

D. interact with signal molecules that diffuse through the plasma membrane.

E. regulate sexual development.

15–7   Intermediate, short answer

Can signaling through the steroid hormone/steroid hormone receptor pathway lead to amplification of the original signal? If so, how?

Nitric Oxide Can Enter Cells to Activate Enzymes Directly (Pages 489–490)

15–8   Easy, multiple choice (Requires information from the whole chapter)

The local mediator nitric oxide stimulates the intracellular enzyme guanylate cyclase:

A. by way of a G-protein-mediated mechanism.

B. by way of a receptor tyrosine kinase.

C. by diffusing into cells and stimulating the cyclase directly.

D. by way of a cell-surface receptor linked to an intracellular signaling pathway.

E. by activation of intracellular protein kinases.

There Are Three Main Classes of Cell-Surface Receptors (Pages 490–491)

15–9   Easy, short answer

Name the three main classes of cell-surface receptors.

15–10   Easy, multiple choice

Which of the following statements are true?

A. All hydrophilic signaling molecules bind to cell-surface receptors that contain at least one membrane-spanning domain.

B. All extracellular signaling molecules are transported across the plasma membrane by their receptor.

C. A cell-surface receptor capable of binding only one natural ligand can mediate only one kind of response.

D. Cells having the same set of receptors can respond in different ways to the same ligand molecules.

E. Foreign substances that occupy receptor sites normally bound by signal molecules always induce the same response that is produced by the natural ligand.

Ion-Channel-linked Receptors Convert Chemical Signals into Electrical Ones (Pages 491–492)

15–11   Intermediate, data interpretation + multiple choice (Requires student also to have studied Chapter 12)

Nerve impulses are triggered when the plasma membrane is depolarized (that is, it becomes less negative) beyond a certain threshhold level. When a nerve cell is treated with neurotransmitters A and/or B, the following results are obtained:

Neurotransmitter added	Impulse triggered?
None	No
A	Yes
B	No
A and B	No

The intracellular and extracellular concentrations of four common ions are given below.

Ion	Intracellular concentration (mM)	Extracellular concentration (mM)
Na^+	10	150
K^+	150	5
Ca^{2+}	10^{-4}	1
Cl^-	5	100

How is the inhibitory neurotransmitter B likely to work?

A. Binds to Na^+ channels and opens them.

B. Binds to K^+ channels and prevents A from opening them.

C. Binds to Ca^{2+} channels and opens them.

D. Binds to Cl^- channels and opens them.

Intracellular Signaling Cascades Act as a Series of Molecular Switches (Pages 492–493)

15–12   Intermediate/easy, multiple choice

Which of the following statements are true?

    A. GTP-binding intracellular signaling molecules are activated by changes in the intracellular concentration of GTP.

    B. All known GTP-binding intracellular signaling molecules are activated on binding GTP.

    C. All known intracellular signaling molecules regulated by phosphorylation are active in the phosphorylated form and inactive in the dephosphorylated form.

    D. All known intracellular signaling molecules regulated by phosphorylation are protein kinases.

    E. All known members of the GTP-binding intracellular signaling molecules are GTPases.

G-PROTEIN-LINKED RECEPTORS (Pages 493–504)

Stimulation of G-Protein-linked Receptors Activates Intracellular G-Protein Subunits (Pages 493–495)

15–13   Easy, multiple choice

When a trimeric G protein is activated by a cell-surface receptor:

    A. the β subunit exchanges its bound GDP for GTP.

    B. the GDP bound to the α subunit is phosphorylated to form bound GTP.

    C. it dissociates into a free β subunit and an αγ subunit.

    D. the α subunit exchanges its bound GDP for GTP.

    E. it dissociates into an active α subunit and an inactive βγ subunit.

15–14   Intermediate, data interpretation + multiple choice

When yeast cells are exposed to mating factor, they stop their progress through the cell cycle and change to the "shmoo" shape in preparation for conjugation. Their response to mating factor is mediated through a receptor linked to a trimeric G protein. The behavior of mutant strains lacking one or more of the subunits of this G protein is summarized below.

Genotype	– Mating factor	+ Mating factor
Wild-type	spherical, dividing	shmoo-shaped, arrested in cell cycle
– α	die as a result of permanent arrest in the cell cycle	die as a result of permanent arrest in the cell cycle
– β	spherical, dividing	spherical, dividing
– γ	spherical, dividing	spherical, dividing
– β, – γ	spherical, dividing	spherical, dividing
– α, – β	spherical, dividing	spherical, dividing
– α, – γ	spherical, dividing	spherical, dividing

Which of the following models are consistent with the behavior of these mutants?

A. α activates the mating response but is inhibited when bound by βγ.

B. βγ activates the mating response but is inhibited when bound by α.

C. The αβγ trimer is inactive; either free α or free βγ is capable of activating the mating response.

D. The αβγ trimer is inactive; both free α and free βγ are required to activate the mating response.

E. The αβγ trimer inhibits the mating response; free α and free βγ have no activity.

Some G Proteins Regulate Ion Channels (Pages 495–496)

15–15   Intermediate, multiple choice

Acetylcholine acts at a G-protein-linked receptor on heart muscle to make the heart beat more slowly by the effect of the G protein on a K$^+$ channel, as shown in Figure Q15–15. Which one or more of the following would enhance this effect of acetylcholine?

A. A high concentration of a non-hydrolyzable analog of GTP.

B. Modification of the acetylcholine-receptor-linked G-protein α subunit by cholera toxin.

C. Mutations in the acetylcholine receptor that weaken the interaction between the receptor and acetylcholine.

D. Mutations in the acetylcholine receptor that weaken the interaction between the receptor and the G protein.

E. Mutations in the G-protein β subunit that weaken the interaction between αβ and the K$^+$ channel.

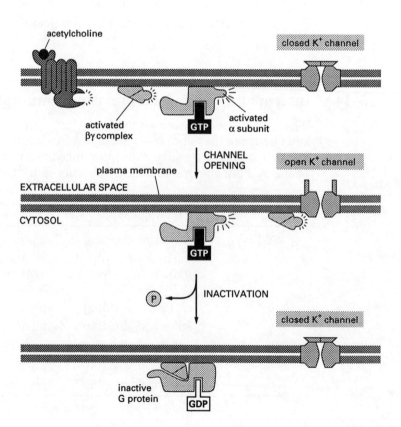

Q15–15

Some G Proteins Activate Membrane-bound Enzymes (Pages 496–497)

15–16   Easy, short answer

(A) Name the two membrane-bound enzymes that are the main targets for activated G proteins.

(B) Which second messengers do each of these enzymes produce?

The Cyclic AMP Pathway Can Activate Enzymes and Turn On Genes (Pages 497–499)

15–17   Easy, matching/fill in blanks (Also requires information from section on pages 501–502)

Match each of the second messengers in the first list with one or more of their targets from the second list, by writing the appropriate letters beside it. A target may be used more than once.

1. Cyclic AMP.   A. Calmodulin.
2. Inositol 1,4,5-trisphosphate.   B. A-kinase.
3. Diacylglycerol.   C. Ca^{2+} channels in ER.
4. Ca^{2+}.   D. C-kinase.

15–18   Intermediate, multiple choice

The cyclic AMP signaling pathway:

   A. involves a series of kinases that become active upon binding cyclic AMP.
   B. is amplified by the catalytic activity of adenylate cyclase.
   C. is triggered in response to changes in intracellular ATP levels.
   D. generally leads to a decrease in the rate of metabolism.
   E. shuts off immediately when cyclic AMP is inactivated by a phosphodiesterase.

The Pathway Through Phospholipase C Results in a Rise in Intracellular Ca^{2+} (Pages 499–501)

15–19   Intermediate/difficult, data interpretation + multiple choice

The differentiation of two related cell lines A and B can be stimulated by a growth factor that binds to a G-protein-linked receptor. This response can be mimicked in cell line A by the addition of ionomycin, a drug that makes membranes permeable to Ca^{2+} ions. Ionomycin does not stimulate the differentiation of cell line B.

(A) Which of the following is the likely plasma membrane target of the G protein activated by the growth factor?

   A. Adenylate cyclase.
   B. Phospholipase C.
   C. K^+ channels.
   D. Ca^{2+} channels.

(B) Which of the following treatments are most likely to stimulate differentiation in ionomycin-treated B cells?

    A. Inositol 1,4,5-trisphosphate.

    B. Phorbol ester, an analogue of diacylglycerol.

    C. EGTA, a compound that binds to free Ca^{2+}.

    D. Antibodies against G-protein α subunits.

    E. Cyclic AMP.

A Ca^{2+} Signal Triggers Many Biological Processes (Pages 501–502)

15–20   Intermediate/difficult, data interpretation + short answer

A calmodulin-regulated kinase (CaM kinase) is involved in learning and memory. This kinase is able to phosphorylate itself, and the kinase activity of the phosphorylated form is independent of the intracellular concentration of Ca^{2+}. Thus the kinase stays active after Ca^{2+} levels have dropped. Mice completely lacking this kinase have severe learning defects but are otherwise normal.

(A) Each of the following mutations also leads to defective memory. For each case explain the reason why.

    A. A mutation that prevents the kinase from binding to ATP.

    B. A deletion of the calmodulin-binding part of the kinase.

    C. A mutation that destroys the site of autophosphorylation.

(B) What might be the effect of a mutation that reduced the strength of binding of the kinase to the phosphatase responsible for inactivating the kinase? Explain your answer.

15–21   Intermediate, short answer (Also requires information from sections on pages 497–501)

In certain pathogenic fungi that invade human cells, the gene regulatory protein Inv activates the transcription of genes for secreted proteases that help the fungi invade. Inv only binds to DNA after having been phosphorylated by a cyclic AMP-dependent protein kinase. The phospholipase C pathway has also been found to influence the activation of Inv, and the cyclic AMP phosphodiesterase from the fungi has been shown to be activated by calmodulin *in vitro*. The cyclic AMP pathway is activated in these fungi by the presence of human extracellular matrix proteins. Assuming that the components of the fungal cyclic AMP and phospholipase C pathways function in the same way as they do in their mammalian counterparts, which of the following will allow transcription of the protease genes in the absence of extracellular matrix proteins?

    A. Inactivation of calmodulin.

    B. Inactivation of the cyclic AMP-dependent protein kinase (A-kinase).

    C. Inactivation of protein kinase C (C-kinase).

    D. Inactivation of phospholipase C.

    E. Inactivation of adenylate cyclase.

Intracellular Signaling Cascades Can Achieve Astonishing Speed, Sensitivity, and Adaptability: Photoreceptors in the Eye (Pages 502–504)

ENZYME-LINKED RECEPTORS (Pages 504–509)
Activated Receptor Tyrosine Kinases Assemble a Complex of Intracellular Signaling Proteins (Pages 505–506)

15–22   Easy, multiple choice (Requires information from the whole chapter)

Which of the following components are found in signaling pathways stimulated by receptor tyrosine kinases?

    A. GTP-binding proteins.

    B. Phosphatases.

    C. Adenylate cyclase.

    D. Ras protein.

    E. Cyclic AMP.

15–23   Easy/intermediate, multiple choice

The growth factor Superchick stimulates proliferation of cultured chicken cells. The receptor that binds Superchick has homology to receptor tyrosine kinases, and many tumor cell lines have mutations in the gene that encodes this receptor. Which of the following types of mutations would be expected to induce uncontrolled proliferation of cells?

    A. A mutation that prevents localization of the receptor to the plasma membrane.

    B. A mutation that prevents dimerization of the receptor.

    C. A mutation that destroys the kinase activity of the receptor.

    D. A mutation that prevents recognition of the receptor by phosphatases.

    E. A mutation that prevents ligand binding.

15–24   Easy/intermediate, multiple choice

The protein Ras:

    A. has one membrane-spanning α helix.

    B. is a protein kinase.

    C. binds directly to phosphorylated tyrosine residues on a variety of receptor proteins.

    D. usually becomes oncogenic when mutated.

    E. is required for both the growth and differentiation of many different types of cells.

Receptor Tyrosine Kinases Activate the GTP-binding Protein Ras (Pages 506–507)

15–25   Difficult, multiple choice + short answer

Male cockroaches with mutations in a receptor tyrosine kinase gene, A, are oblivious to the charms of their female comrades. This particular receptor tyrosine kinase binds to a small molecule secreted by sexually mature females. Most males carrying mutations in the gene for Ras protein are also unable to respond to females. You have just read a paper in which the authors describe how they have screened cockroaches with the defective kinase A gene for additional mutations that partially restore the ability of males to respond to females. These mutants have a defect in a gene that the authors call C. Which of the following proteins could be encoded by gene C? Explain your answer.

   A. A protein that stimulates exchange of GDP for GTP by the Ras protein.
   B. A protein that stimulates hydrolysis of GTP by the Ras protein.
   C. A protein kinase activated by the Ras protein.
   D. An adaptive protein that mediates the binding of the receptor A to the Ras protein.
   E. A transcription factor required for the expression of the Ras gene.

Protein Kinase Networks Integrate Information to Control Complex Cell Behaviors (Pages 508–509)

15–26   Easy, multiple choice

A kinase will act as an integrating device in signaling if it:

   A. phosphorylates more than one substrate.
   B. catalyzes its own phosphorylation.
   C. binds to more than one inhibitor or activator.
   D. has at least two sites at which the kinase itself can be phosphorylated by different kinases.
   E. initiates phosphorylation cascades.

Answers

A15–1.   A. Hormones are made by <u>endocrine</u> cells and are carried to their target cells by the <u>bloodstream</u>.

B. Paracrine signaling involves signal molecules known as <u>local mediators</u>.

C. Neuronal signaling is a specialized example of <u>paracrine signaling</u>.

D. If a cell cannot respond to a particular signal molecule it may be because it does not carry a <u>receptor</u> for that molecule.

E. In cell signaling, the conversion of the signal from one form into another is known as signal <u>transduction</u>.

A15–2.   D. Hormones are secreted into the circulatory system and are diluted in the bloodstream; therefore they need to be produced in fairly large quantities and have a relatively high affinity for their receptors. Hormones also need to be relatively stable in order to survive the journey through the bloodstream to the target. Paracrine signals diffuse a short distance before they reach their target and thus are not diluted much during their journey from the signaling cell to the target cell. Both hormones and paracrine signals are soluble in water. Hormones are only produced by endocrine cells.

A15–3.   A and C. Since cell line A proliferates only in the presence of F and since cell line B can proliferate in the absence of F once R is expressed, binding of factor F must be required for proliferation of cell line A but not cell line B. Because expression of R is sufficient to induce proliferation of cell line B in the absence of F, there is some other factor, X, presumably in the serum, that is capable of binding to R and stimulating B. Since R binds to X, it probably does not bind to F, since most receptors do not bind more than one ligand. We do not know the effect of eliminating receptor R from cell line A; therefore, we cannot say if binding of receptor R to its ligand is required for proliferation cell line A. Since R is not the receptor for F, we cannot say that cell line B does not contain a receptor for F. Cell line B could have a receptor for F and respond to F in ways that have nothing to do with cell proliferation.

A15–4.   C and E. Each activated adenylate cyclase molecule generates many molecules of cyclic AMP. Each activated A-kinase molecule can phosphorylate many molecules of target proteins. However, each signal molecule activates only one receptor, each activated G protein activates only one molecule of cyclase, and one cyclic AMP molecule can help activate only one A-kinase molecule.

A15–5.   E. The more complex pathway increases the opportunity for the response to be regulated effectively with respect to conditions inside and outside the cell, since each intermediate step represents a step that can be enhanced or inhibited. The more complex pathway, however, is slower (A) and in this example does not move the signal away from the membrane (B). Since there is generally a one-to-one interaction between each component of the pathway in this case, the more complex pathway also does not amplify or distribute the signal (C and D), although if Q stays on its receptor long enough, each receptor could interact with more than one αβ complex, thereby amplifying the response.

A15–6.   D. All members of the steroid hormone receptor family are intracellular proteins (thus A is not correct) that interact with molecules that can diffuse through the plasma membrane unaided. Steroid hormone receptors can be cytoplasmic or nuclear (C). They have no enzymatic activity (B), are activated by either steroids or other hydrophobic signaling moelcules and affect a wide variety of processes, including, but not limited to, sexual development (E).

A15–7.   Since the interactions of the signal molecule with its receptor and of the activated receptor with its gene are both one-to-one, there is no amplification in this part of the signaling pathway. The signal can, however, be amplified when the target gene is transcribed, since multiple copies of mRNA are usually produced from a gene once it has been switched on, and multiple copies of protein can be made from each mRNA molecule.

A15–8.   C.

A15–9.   Ion-channel-linked receptors; G-protein-linked receptors; enzyme-linked receptors.

A15–10. A and D. A hydrophilic molecule cannot diffuse across the membrane and can therefore only affect a cell if it binds to a cell-surface receptor that spans the bilayer (A). Cells having the same set of receptors can respond in different ways to the same ligand molecules, since the response to a receptor also depends on the types of intracellular relay pathways present in the cell (D). Most signal molecules remain outside of the cell bound to the extracellular domain of the receptor, while the intracellular domain mediates signal transduction; although many signal molecules are endocytosed with their receptor, they remain inside membrane-bounded compartments and are therefore not transported across a membrane (B). A cell-surface receptor capable of binding only one natural ligand can mediate more than one kind of response, depending on a variety of conditions, such as the types of intracellular signaling molecules present in the cell type on which the receptor is being expressed (C). Foreign substances that occupy receptor sites normally bound by extracellular signal molecules can sometimes induce the same response that is produced by the natural ligand, but they often block the binding of the natural ligand without stimulating an intracellular signal (E).

A15–11. D. Excitatory neurotransmitters such as A trigger a nerve impulse by opening cation channels, which allows Na^+ to enter the cell down its steep electrochemical gradient; this depolarizes the plasma membrane. Inhibitory neurotransmitters such as B usually open Cl^- channels, which allows Cl^- to enter the cell when the membrane begins to depolarize; this tends to make it harder to depolarize the membrane further, thereby inhibiting the action of A.

A15–12. B and E. Activation of the GTP-binding class of intracellular signaling molecules is not dependent on fluctuations in the intracellular concentration of GTP (which is fairly constant) (A); instead, signaling is regulated by proteins which stimulate GTP hydrolysis or GDP/GTP exchange. Some intracellular signaling molecules regulated by phosphorylation are active in the dephosphorylated form and inactive in the phosphorylated form (C). Many, but not all, members of the class of intracellular signaling molecules regulated by phosphorylation are kinases (D).

A15–13. D. E is incorrect since the βγ subunit may have an effect instead of or as well as the α subunit. The other statements are simply untrue.

A15–14. B. If βγ activates the mating response but is inhibited when bound by α, then β- and γ- mutants should be viable but unable to respond to mating factor, while α- mutants will activate the mating response in the absence of mating factor and will therefore become permanently arrested in the cell cycle and die. All of these predictions are borne out by the data. If α activated the mating response but was inhibited when bound by βγ, α- mutants would be unable to respond to mating factor, and γ- or β- mutants would be dead. If the αβγ trimer were inactive and either free α or free βγ were capable of activating the mating response, then β-, γ- and α- mutants would be dead. If the αβγ trimer were quiescent and both free α and free βγ were required to activate the mating response, then β-, γ-, and α- mutants would all be unable to respond to mating factor. If the αβγ trimer inhibited the mating response and free α and free βγ had no activity, then β-, γ-, and α- mutants would all be dead.

A15–15.   A and B. The heart is induced to beat more slowly by binding of acetylcholine to a G-protein-linked receptor whose activated βγ subunit binds to and opens K+ channels. High concentrations of a non-hydrolyzable analogue of GTP or modification of the acetylcholine-receptor-linked G-protein α subunit by cholera toxin (which also prevents GTP hydrolysis) will increase the length of time that the G-protein βγ subunit remains free of α and able to activate the K+ channel; they will therefore enhance the effect of acetylcholine. Receptor mutations that weaken its interaction with acetylcholine, receptor mutations that weaken its interaction with the G protein, or β mutations that weaken the interaction between αβ and the K+ channel, all make it more difficult for the signal to proceed from the receptor to the K+ channel.

A15–16.   (A) Adenylate cyclase and phospholipase C. (B) Adenylate cyclase produces cyclic AMP. Phospholipase C produces inositol 1,4,5-trisphosphate and diacylglycerol.

A15–17.   1, B; 2, C; 3, D; 4, A and D.

A15–18.   B is correct. A is incorrect, as only one kinase, the A-kinase, is activated by binding cyclic AMP. C, D, and E are incorrect, as the cyclic AMP pathway is triggered by activating adenylate cyclase in response to G-protein-linked receptors, generally leads to an increase in the rate of metabolism, and can be shut off only after cyclic AMP levels have dropped and all proteins phosphorylated by A-kinase have been dephosphorylated.

A15–19.   (A) B. (B) B. Ionomycin increases the cytoplasmic concentration of Ca^{2+} by allowing Ca^{2+} from the surroundings and the endoplasmic reticulum to rush into the cytoplasm. Since ionomycin can mimic the effect of a growth factor that binds to a G-protein-linked receptor, at least in A cells, the growth factor probably acts by activating phospholipase C, which has two effects: it increases intracellular Ca^{2+} concentrations and increases the activity of C-kinase. To induce cells to differentiate, then, we would want to increase both Ca^{2+} levels and C-kinase activity. In ionomycin-treated cells, Ca^{2+}-stimulated signaling molecules will already be activated and substances which act by increasing intracellular Ca^{2+} levels (such as inositol 1,4,5-trisphosphate) will not have much additional effect. Phorbol ester, on the other hand, will stimulate C-kinase and enhance the effect of ionomycin. EGTA will lower the concentration of Ca^{2+}, counteracting the effect of ionomycin. Antibodies against the G-protein α subunit will decrease its ability to activate phospholipase C and will oppose the effects of ionomycin. Stimulation of A-kinase may or may not affect differentiation; since we are pretty sure that C-kinase is involved in differentiation, phorbol ester is more likely than cyclic AMP to enhance the effect of ionomycin in B cells.

A15–20.   (A) Since a complete lack of the CaM kinase causes a loss of memory, we can assume that mutations that lead to inactivation of the kinase would also have a deleterious effect on memory. A, Protein kinases have a binding site for ATP, which is the source of the phosphate used for phosphorylation; if the kinase cannot bind ATP it will be inactive. B, Since CaM kinases are activated by binding to calmodulin in the presence of Ca^{2+}, deletion of the calmodulin-binding portion would inactivate the kinase. C, A mutation that destroys the site of autophosphorylation will also impair the function of the kinase, since the kinase will become inactive as soon as Ca^{2+} levels drop. (B) Mutations that make the kinase less likely to bind the phosphatase will increase the time that the kinase remains active following a transient increase in Ca^{2+} levels. This could lead to increased learning ability.

A15–21.   A and D. Since Inv is activated by A-kinase, any condition that increases the cellular concentration of cyclic AMP could increase transcription of the Inv-activated protease genes. Since calmodulin binds to and activates cyclic AMP phosphodiesterase *in vitro*, inactivation of calmodulin will likely increase the amount of cyclic AMP. Likewise, inactivation of phospholipase C would decrease the levels of inositol 1,4,5-trisphosphate, which would decrease the levels of Ca^{2+} and decrease the activity of calmodulin. Inactivation of A-kinase or adenylate cyclase would abolish phosphorylation of Inv and hence would wipe out Inv-activated transcription. Inactivation of C-kinase would not affect the levels of either Ca^{2+} or cyclic AMP and therefore should not affect phosphorylation of Inv.

A15–22.   A, B, and D.

A15–23.   D. Tyrosine receptor kinases are activated by ligand-induced dimerization, which allows the receptors to phosphorylate themselves and activate intracellular signaling proteins that are stimulated by the phosphorylated receptor. After it is activated, the receptor is dephosphorylated, and thereby inactivated, by a phosphatase. Therefore, a mutation in the phosphatase will inappropriately increase the activity of the receptor and lead to uncontrolled cell proliferation. Mutations that prevent localization of the receptor to the plasma membrane, dimerization of the receptor (including mutations that prevent ligand binding), or autophosphorylation will inactivate the receptor.

A15–24.   E. Ras is not a membrane-spanning protein. It has GTPase but not kinase activity. It binds directly to Ras-activating proteins (which bind to adapter proteins that bind to phosphorylated tyrosine residues on receptor proteins). It becomes oncogenic only when mutated in a way that permanently switches the protein on; most mutations will simply inactivate the protein and are not oncongenic.

A15–25.   B. Mutations that increase the activity of Ras should mimic the effect of stimulating the receptor A in a receptor-independent fashion. Since the intracellular concentration of GTP is higher than that of GDP, some proportion of the Ras molecules is expected to be GTP-bound and active; ridding the cells of a protein which stimulates GTP hydrolysis will increase this pool of active Ras. Mutants that cannot stimulate exchange of GDP for GTP by Ras will have the same phenotype as mutants lacking Ras, as will mutants lacking a transcription factor required for expression of the Ras gene. Mutants lacking the kinase activated by Ras will be unable to transmit any signal from Ras onward. Defects in an adaptive protein that mediates the binding of receptor A to Ras will have no further effect on a mutant already lacking the receptor.

A15–26.   C and D. Integrating devices are able to relay signals from more than one signaling pathway. Binding to more than one inhibitor or activator or having sites that can be phosphorylated by different kinases allows a kinase (or any other signaling molecule) to be affected by more than one upstream signal. A, B, and E affect the output signal that a kinase is able to produce, not its ability to integrate incoming signals.

16 Cytoskeleton

Questions

16–1   Easy, matching/fill in blanks

For each of the following sentences, fill in the blanks with the correct word selected from the list below. Use each word only once.

    A. The protein _____ is the subunit of microtubules.

    B. The protein _____ is the subunit of microfilaments.

    C. Intermediate filaments are _____ and _____.

    D. Microtubules are _____ and _____.

    E. Actin filaments are _____ and _____.

hollow; lamin; rigid; actin; strong; tubulin; thin; ropelike; keratin; flexible.

16–2   Easy, art labeling

Identify the cytoskeletal structures in the epithelial cells shown in Figure Q16–2.

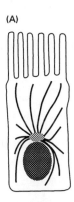 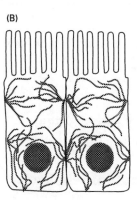 

Q16–2

25 μm

INTERMEDIATE FILAMENTS (Pages 514–518)
Intermediate Filaments Are Strong and Durable (Page 515–516)
Intermediate Filaments Strengthen Cells Against Mechanical Stress (Pages 516–518)

16–3   Intermediate, multiple choice + short answer (Requires information from the whole chapter)

(A) What properties of intermediate filament monomers distinguish them from the monomers that make up actin filaments or microtubules?

    A. They bind to each other covalently.

    B. They are fibrous rather than globular proteins.

    C. They do not bind and hydrolyze nucleotides.

D. There are numerous different types in different cell types.

E. They are glycosylated.

(B) Which of the above differences makes intermediate filaments less dynamic structures than actin filaments or microtubules?

16–4   Easy, multiple choice

Intermediate filaments:

A. are always polymers of keratin and vimentin.

B. bear mechanical stress well.

C. indirectly link neighboring epithelial cells.

D. are specific to epithelial cells.

E. are found in the nucleus.

16–5   Intermediate, short answer

(A) What types of covalent modification regulate lamin assembly and disassembly during mitosis?

(B) If you could prevent these modifications from occurring, what would you expect to happen?

MICROTUBULES (Pages 518–529)
Microtubules Are Hollow Tubes with Structurally Distinct Ends (Page 519)

16–6   Easy, multiple choice

Microtubules are structural components in which of the following?

A. Cilia.

B. Centrioles.

C. Mitotic spindle.

D. Bacterial flagella.

E. Nuclear lamina.

16–7   Easy, short answer

Place the following in order of size, from the smallest to the largest.

A. Protofilament.

B. Microtubule.

C. α-tubulin.

D. Tubulin dimer.

E. Mitotic spindle.

16–8   Easy, art labeling (Requires information from Figure 16–10 on page 521)

In the three cell outlines in Figure Q16–8 indicate how the microtubules will be arrranged, showing clearly their free and attached ends. On each figure indicate the plus end and the minus end for one of the microtubules.

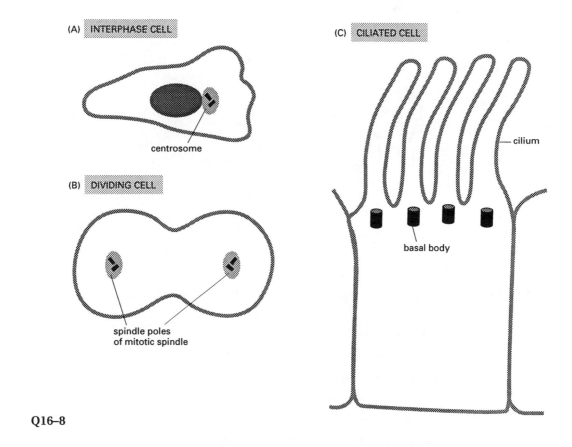

Q16–8

16–9   Intermediate/difficult, short answer (Requires information from the whole chapter)

(A) What structural feature of a microtubule gives it a structural polarity?

(B) Why is this polarity important in microtubule function?

Microtubules Are Maintained by a Balance of Assembly and Disassembly (Pages 519–521)

16–10   Intermediate, short answer (Requires information from Chapter 8)

You are interested in looking at the chromosomes of a particular cell line under the microscope. Why might you decide to treat your cell culture with the drug colchicine before looking at them?

16–11   Intermediate, data interpretation

The graph in Figure Q16–11 shows the time-course of the polymerization of pure tubulin *in vitro*. You can assume that the starting concentration of free tubulin is much higher than it is in cells.

(A) Explain the reason for the initial lag in the rate of microtubule formation.

(B) Why does the curve level out after point C?

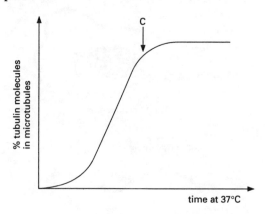

Q16–11

The Centrosome Is the Major Microtubule-organizing Center in Animal Cells

16–12   Easy, multiple choice

The minus end of a microtubule:

    A. is the end that is embedded in the centrosome.

    B. is the faster growing end.

    C. often has a "GTP cap."

    D. is bound to γ-tubulin rings in the centrosome.

    E. is where polymerization occurs fastest for a free microtubule.

16–13   Easy, multiple choice

The hydrolysis of GTP to GDP carried out by tubulin molecules:

    A. provides the energy needed for tubulin to polymerize.

    B. occurs because the pool of free GDP has run out.

    C. tips the balance in favor of microtubule assembly.

    D. is prevented by attachment of the plus end of the microtubule to an organelle.

    E. allows the behavior of microtubules called dynamic instability.

16–14   Intermediate, multiple choice (Note to instructors: should not be used on the same test paper as Question 16–9)

A microtubule is said to have a polarity, with a plus end and a minus end, because:

    A. one end of the microtubule is more highly charged than the other end.

    B. one end of the microtubule has a smaller diameter than the other end.

    C. the tubulin subunits in each protofilament are all lined up in the same direction, head to tail, leaving the "head" end of tubulin subunits exposed at one end of the microtubule and the "tail" end of tubulin subunits exposed at the other end.

    D. one end of the microtubule is more soluble in water than the other end.

16–15   Easy, multiple choice

The microtubules in a cell form a structural framework that can have all the following functions except:

    A. holding internal organelles such as the Golgi apparatus in particular positions in the cell.

    B. creating long thin cytoplasmic extensions that protrude from one side of the cell.

    C. strengthening the plasma membrane.

    D. moving materials from one place to another inside a cell.

Motor Proteins Drive Intracellular Transport (Pages 525–526)

16–16   Easy, multiple choice

Kinesins and dyneins:

    A. are inhibited by colchicine and taxol.

    B. move along both microtubules and actin filaments.

    C. often move in opposite directions to each other.

    D. derive their energy from GTP hydrolysis.

    E. have tails that bind to the filaments.

Organelles Move Along Microtubules (Pages 526–527)

16–17   Easy, art labeling

Match the following labels to the numbered label lines on Figure Q16–17.

    A. Dynein.

    B. Kinesin.

    C. ATP-binding head of motor protein.

    D. Cargo.

    E. Minus end of microtubule.

    F. Tail of motor protein.

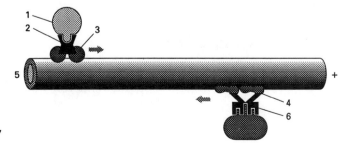

Q16–17

Cilia and Flagella Contain Stable Microtubules Moved by Dynein (Pages 527–529)

16–18   Easy, matching/fill in blanks (Requires information from all sections on pages 514–529)

For each of the terms in the following list, indicate whether they refer to intermediate filaments (IF) and their function, or to microtubules (M) and their function.

    A. 9+2 array.

    B. Basal body.

    C. Centrosome.

    D. Cilia.

    E. Dynein.

    F. GTP cap.

    G. Keratin.

    H. Kinesin.

    I. Mitotic spindle.

    J. Neurofilaments.

    K. Nuclear lamina.

    L. Protofilaments.

    M. Tubulin.

16–19   Intermediate/difficult, short answer

Consider Figure Q16–19, which shows two isolated outer doublet microtubules from a eucaryotic flagellum with their associated dynein molecules.

(A) Sketch what will happen to this structure if it is supplied with ATP.

(B) Sketch what will happen to this structure if the linking proteins are removed and it is supplied with ATP.

(C) In a complete flagellum, what would happen if all the dynein molecules were active at the same time?

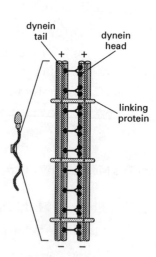

Q16–19

ACTIN FILAMENTS (Pages 529–542)

Actin Filaments Are Thin and Flexible (Pages 530–531)

16–20   Easy, matching/fill in blanks

Fill in the blanks in the following sentences.

"The monomer subunit of microfilaments is the protein _____. Chains of monomers form a _____-stranded helix, which has a structural _____, with a _____ end and a _____ end."

Actin and Tubulin Polymerize by Similar Mechanisms (Pages 531–532)

16–21   Intermediate/difficult, short answer

(A) What biochemical reaction contributes to dynamic instability of actin filaments?

(B) Explain briefly how it does so.

(C) ATPγS is a synthetic chemical very similar to ATP, except that it cannot be hydrolyzed. What would you expect to happen if you substituted it for ATP in a polymerizing mixture of actin and actin filaments?

Many Proteins Bind to Actin and Modify Its Properties (Pages 532–533)

16–22   Easy, multiple choice

Profilin and thymosin:

A. bind to actin filaments.

B. crosslink actin in the cell cortex.

C. bind to actin monomers.

D. help to regulate actin polymerization.

E. serve as tracks for the movement of organelles.

An Actin-rich Cortex Underlies the Plasma Membrane of Most Eucaryotic Cells (Page 533)

16–23   Easy, multiple choice

Polymerization and depolymerization of actin are essential for:

A. production of contractile forces in cells.

B. movement of cilia and flagella.

C. crawling movements of cells over surfaces.

D. protrusive movements of the cell surface.

E. maintaining the shape of a red blood cell.

Cell Crawling Depends on Actin (Pages 533–536)

16–24   Intermediate, short answer

How are lamellipodia and filopodia formed?

Actin Associates with Myosin to Form Contractile Structures (Pages 536–537)

16–25   Easy, multiple choice

Myosins:

A. bind and hydrolyze GTP.

B. form contractile rings with actin in dividing cells.

C. polymerize in a double-stranded helix.

D. have heads which contact actin filaments.

E. move towards the minus end of an actin filament.

16–26   Intermediate, short answer

Cells contain many diverse types of myosins. What two structural/functional features do they all have in common?

During Muscle Contraction Actin Filaments Slide Against Myosin Filaments (Pages 538–539)

16–27   Intermediate, multiple choice

During muscle contraction:

A. the myosin filaments shorten, bringing the actin filaments closer together.

B. the thin filaments are bound to proteins in the Z discs.

C. ATP hydrolysis by actin monomers drives the contraction.

D. the myosin head "walks" along the actin filament via conformational changes as a consequence of the binding and hydrolysis of ATP.

E. the myosin filament moves from the minus end of the actin filament toward the plus end.

Muscle Contraction Is Triggered by a Sudden Rise in Ca^{2+} (Pages 539–542)

16–28   Easy, short answer (Requires information from Chapter 12)

In contraction of skeletal muscle:

(A) how does depolarization of the muscle cell membrane cause an increase in the concentration of Ca^{2+} in the cytoplasm of the muscle cell?

(B) what would be the effect of diminishing ATP levels on muscle contraction?

(C) how does the increase of cytoplasmic Ca^{2+} trigger the muscle cell to contract?

Answers

A16–1.   A. The protein <u>tubulin</u> is the subunit of microtubules.

B. The protein <u>actin</u> is the subunit of microfilaments.

C. Intermediate filaments are <u>strong</u> and <u>ropelike</u>.

D. Microtubules are <u>hollow</u> and <u>rigid</u>.

E. Actin filaments are <u>thin</u> and <u>flexible</u>.

A16–2.   A = microtubules; B = intermediate filaments; C = actin filaments (microfilaments).

A16–3.   (A) B, C, and D. (B) C.

A16–4.   B, C, and E.

A16–5.   (A) Phosphorylation causes disassembly. Dephosphorylation causes reassembly. (B) If phosphorylation were prevented, the intermediate filaments would not fall apart, the nuclear lamina would remain intact, and the nucleus would not be able to divide. If dephosphorylation were prevented, the lamin subunits, and hence the nuclear envelope, would not reassemble at the end of mitosis.

A16–6.   A, B, and C.

A16–7.   C, D, A, B, E.

A16–8.   Figure A16–8.

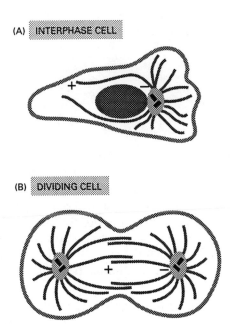

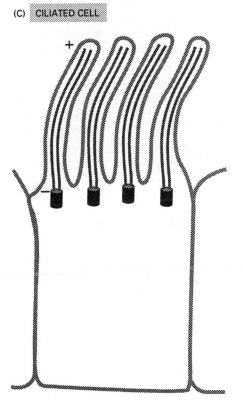

A16–8

A16–9.   (A) Microtubules have an inherent polarity because their α/β tubulin subunits are all arranged with the same orientation in the polymer with β tubulin exposed at one end of the microtubule and α tubulin exposed at the other. (B) The motor proteins that interact with microtubules travel in one direction—either from the + end to the – end or vice-versa—along the microtubule. Thus microtubules can mediate movement in a specified direction.

A16–10.   In order to see chromosomes under the microscope you need cells that are in mitosis, when the chromosomes are condensed and therefore visible. In a normal population of dividing cells only a few cells will be in mitosis at any one time, and they can be hard to find. Colchicine will arrest all the cells in your sample in the middle of mitosis, a point at which the chromosomes are all condensed and readily visible.

A16–11.   (A) Before they can polymerize to form microtubules, tubulin molecules have to form small aggregates that act as nucleation centers. This aggregation step is slow as the molecules have to come together in the right configuration. This is why there is a lag phase before microtubules start to be formed. (B) After point C an equilibrium point has been reached where the rates of polymerization and depolymerization are exactly balanced.

A16–12.   A and D.

A16–13.   E. The hydrolysis of GTP to GDP occurs after a GTP-bound tubulin molecule is incorporated into a microtubule, and it makes the microtubule more susceptible to disassembly. It is the resulting switch in microtubule stability that gives rise to the phenomenon known as dynamic instability.

A16–14.   C. In each protofilament, the α tubulin monomer in each dimeric tubulin subunit is bound to the β tubulin monomer in the next tubulin subunit; moreover, all of the 13 protofilaments are lined up in the same direction. The word "polar" has other meanings in biology, and thus the confusion in A and D, which are both incorrect.

A16–15.   C. One function of actin filaments but not microtubules is to provide a meshwork beneath the plasma membrane that helps to form and strengthen this membrane. Microtubules have all of the other functions that are described.

A16–16.   C.

A16–17.   A, 6; B, 3; C, 4; D, 1; E, 5; F, 2.

A16–18.   M = A, B, C, D, E, F, H, I, L, M. IF = G, J, K.

A16–19.   (A) Figure A16–19A (*next page*). (B) Figure A16–19B (*next page*). (Note to instructor: this question should be marked as correct only if the microtubule is shown bending in the correct direction (in A) and the correct microtubule is shown pushed forward (in B)). (C) The flagellum will not bend because there is no significant relative motion of one microtubule doublet to another, since each is trying to push its neighbor forward at the same time. For the flagellum to bend, a few dynein molecules on one side of the flagellum must be selectively activated.

A16–20.   The monomer subunit of microfilaments is the protein <u>actin</u>. Chains of monomers form a <u>two</u>-stranded helix which has a structural <u>polarity</u>, with a <u>plus</u> end and a <u>minus</u> end.

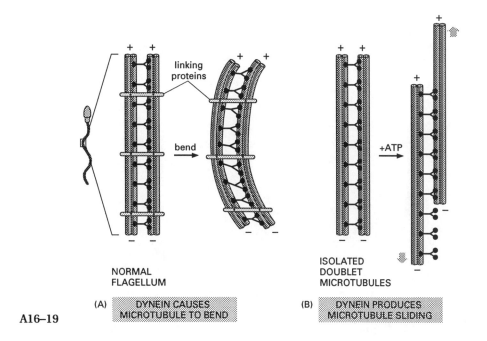

A16–19

A16–21.   (A) The hydrolysis of ATP to ADP. (B) Each free actin monomer carries a tightly bound ATP
            molecule, which is hydrolyzed to ADP soon after incorporation of the actin monomer into the
            filament. Hydrolysis of the bound ATP reduces the strength of binding between the monomers
            and decreases the stability of the actin polymer, thus promoting depolymerization. (C) The
            actin filaments would grow rapidly and steadily until the pool of free actin monomers and
            ATPγS was almost all used up; finally a chemical equilibrium would be reached where the (low)
            rate of dissociation of actin-ATPγS from the exisiting filaments balanced the (low) rate of addi-
            tion of actin-ATPγS to them.

A16–22.   C and D.

A16–23.   C and D.

A16–24.   These structures are formed by the rapid growth of cortical actin filaments that connect at
            their plus ends to nucleation centers in the plasma membrane and which grow at the point of
            this attachment. This rapidly pushes the plasma membrane outward.

A16–25.   B and D.

A16–26.   1. They all have a site at which ATP is bound and hydrolyzed, i.e., they all have ATPase activity.
            2. They all have a site at which they can bind to an actin filament.

A16–27.   B, D, and E.

A16–28.   (A) The action potential produced as a result of depolarization of the muscle cell membrane
            spreads along the T-tubules and causes voltage-gated Ca^{2+} channels in the adjacent sarcoplas-
            mic reticulum (SR) membrane to open. This releases Ca^{2+} from the SR, where it is stored at
            high concentration, into the muscle cell cytosol. (B) Muscle contractions would become slower
            and weaker. (C) The actin filaments in skeletal muscle are associated with troponin C and
            tropomyosin. When Ca^{2+} binds to troponin C, the troponin C undergoes a change in shape.
            This displaces the attached tropomyosin from its position on the actin filament, uncovering a
            myosin-binding site. Myosin can then bind to the actin and produce a contraction.

17 Cell Division

Questions

OVERVIEW OF THE CELL CYCLE (Pages 549–551)
The Eucaryotic Cell Cycle Is Divided into Four Phases (Pages 549–550)

17–1 Easy, matching/fill in blanks

For each of the following sentences, fill in the blanks with the correct terms selected from the list below. Use each term only once.

"In the eucaryotic cell cycle, DNA replication occurs during _____, nuclear and cytoplasmic division occur during _____, and cell growth occurs during _____. In a typical somatic cell, _____ is the longest phase of the cell cycle."

interphase (G_1 + S + G_2); G_1; G_2; S phase; M phase.

17–2 Easy, short answer

What would be the most obvious outcome of repeated cell cycles consisting of S phase and M phase only?

17–3 Intermediate, multiple choice

A mutant yeast strain stops growing and dividing when shifted from 25°C to 37°C. When this strain is analyzed, at either 25°C or 37°C, using a machine that sorts cells according to the amount of DNA they contain, the following graphs are obtained (Figure Q17–3).

Which of the following would explain the behavior of your mutant?

A. Inability to initiate DNA replication.

B. Defect in chromosome condensation.

C. Defect in centrosome duplication.

D. Defect in cytokinesis.

E. Inappropriate production of a signal that causes the cell to remain in G_1.

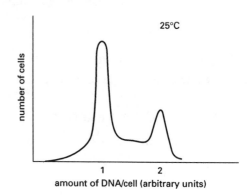

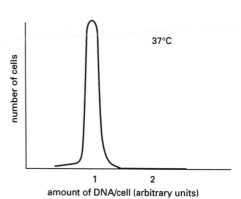

Q17–3

The Cytoskeleton Carries Out Both Mitosis and Cytokinesis (Page 551)

17–4   Intermediate, multiple choice

Which of the following statements are true?

 A. The mitotic spindle is composed mainly of actin filaments.

 B. The mitotic spindle assembles during late mitosis.

 C. Drugs that cause tubulin depolymerization will inhibit progression through the cell cycle in eucaryote and procaryote cells.

 D. Cytokinesis is achieved by a ring of microtubules that forms on the inner surface of the plasma membrane.

 E. The contractile ring is formed only in animal cells.

Some Organelles Fragment at Mitosis (Page 551)

17–5   Easy, multiple choice

Which of the following structures break into fragments prior to cell division?

 A. Mitochondria.

 B. Nuclear membrane.

 C. Centrosomes.

 D. Lysosomes.

 E. Ribosomes.

 F. Chloroplasts.

 G. Endoplasmic reticulum.

MITOSIS (Pages 552–560)
The Mitotic Spindle Starts to Assemble in Prophase (Pages 552–553)

17–6   Easy, short answer

List the six stages of M phase of the cell cycle in the order in which they occur.

17–7   Easy, art labeling (Requires information from Panel 17–1)

Which stage of mitosis in an animal cell does each part of Figure Q17–7 (*next page*) represent?

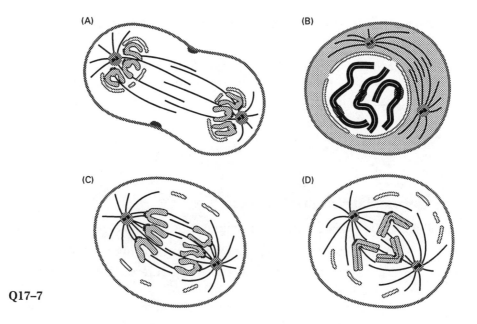

Q17–7

17–8   Intermediate, multiple choice

In which of the following ways does the cytoskeleton of an animal cell change between G_1 and prophase of the cell cycle?

A. The phosphorylation state of centrosomal proteins changes.

B. The number of centrosomes increases.

C. The number of microtubules per centrosome decreases.

D. Some microtubules become stabilized by interacting with each other.

E. The average length of centrosomal microtubules increases.

Chromosomes Attach to the Mitotic Spindle at Prometaphase (Pages 553–557)

17–9   Intermediate, multiple choice

Disassembly of the nuclear envelope:

A. occurs during prophase.

B. involves the separation of the inner nuclear membrane from the outer nuclear membrane.

C. results in the conversion of the nuclear envelope into protein-free membrane vesicles.

D. is triggered by the phosphorylation of integrins.

E. must occur in order for kinetochore microtubules to form.

17–10   Intermediate, short answer (Requires information from sections on pages 563–567)

(A) In a cell that has eight chromosomes (four pairs of homologous chromosomes) in G_1 phase, how many kinetochores will be present in the cell at mitotic prophase?

(B) How many kinetochores will be present in the daughter of such a cell at prophase of the second meiotic division?

17–11   Intermediate, multiple choice

Which of the following statements about kinetochores are true?

A. A kinetochore often binds to more than one microtubule.

B. They contain DNA-binding proteins that recognize sequences at the telomere of the chromosome.

C. Kinetochore proteins bind to the tubulin molecules at the very tip of the plus end of microtubules.

D. Kinetochores assemble on chromosomes that lack centromeres.

E. Kinetochores bind the polar microtubules, converting them to kinetochore microtubules.

Chromosomes Line Up at the Spindle Equator at Metaphase (Page 557)

17–12   Intermediate, multiple choice

A friend declares that chromosomes are held at the metaphase plate by microtubules that push on each chromosome from opposite sides. Which of the following observations best supports your belief that microtubules are pulling on the chromosomes, not pushing them?

A. The jiggling movement of chromosomes at the metaphase plate.

B. The way that chromosomes behave when the attachment between sister chromatids is severed.

C. The way that chromosomes behave when the attachment to one kinetochore is severed.

D. The shape of chromosomes as they move toward the spindle poles at anaphase.

E. The behavior of the spindle when colchicine is added to cells.

Daughter Chromosomes Segregate at Anaphase (Pages 557–559)

17–13   Intermediate, multiple choice

Which of the following statements are true?

A. Anaphase is triggered by the phosphorylation of proteins that hold the sister chromatids together.

B. Anaphase A must be completed before anaphase B can take place.

C. In cells where anaphase B predominates, the spindle will elongate much less than in cells where anaphase A dominates.

D. In anaphase A, both kinetochore and polar microtubules shorten.

E. In anaphase B, microtubules associated with the cell cortex shorten.

17–14   Difficult, multiple choice + short answer

Which of the following would be expected to affect only anaphase B? Explain your answer.

A. An antibody against myosin.

B. ATPγS, a nonhydrolyzable ATP analog that binds to and inhibits ATPases.

C. An antibody against motor proteins that move from the plus end of micro-tubules to the minus end.

D. An antibody against motor proteins that move from the minus end of micro-tubules towards the plus end.

E. Colchicine.

The Nuclear Envelope Re-forms at Telophase (Pages 559–560)

17–15   Intermediate, multiple choice (Requires information from section on pages 562–563)

Which of the following precede re-formation of the nuclear envelope during M phase?

A. Formation of the contractile ring.

B. Decondensation of chromosomes.

C. Reassembly of the nuclear lamina.

D. Formation of the phragmoplast.

E. Transcription from chromosomes.

17–16   Intermediate, multiple choice

A serine to alanine mutation in the phosphorylation site of a lamin protein will:

A. cause cells to arrest at telophase.

B. stabilize the nuclear lamina and thereby prevent its disassembly.

C. prevent nuclear pore assembly.

D. prevent mitotic spindle formation.

E. prevent chromosome condensation.

CYTOKINESIS (Pages 560–563)
The Mitotic Spindle Determines the Plane of Cytoplasmic Cleavage (Pages 560–561)

17–17   Intermediate, multiple choice

The cleavage furrow:

A. is a puckering of the plasma membrane caused by narrowing of a ring of fila-ments attached to the membrane.

B. begins to form at the end of telophase.

C. will not begin to form in the absence of a mitotic spindle.

D. always forms perpendicular to the polar microtubules.

E. always forms in the middle of the cell.

The Contractile Ring of Animal Cells Is Made of Actin and Myosin (Pages 561–562)

17–18   Intermediate, multiple choice

Cytokinesis in animal cells:

A. requires ATP.

B. leaves a small circular scar of actin filaments on the inner surface of the plasma membrane.

C. is often followed by phosphorylation of integrins in the cell's plasma membrane.

D. is generally accompanied by rearrangement of filaments making up the cell cortex.

E. is assisted by motor proteins that pull on microtubules attached to the cell cortex.

Cytokinesis in Plant Cells Involves New Cell-Wall Formation (Pages 562–563)

17–19   Intermediate, multiple choice

Cytokinesis in plant cells:

A. does not involve a contractile ring.

B. does not require ATP.

C. always occurs perpendicular to the long axis of the mitotic spindle.

D. involves formation of the phragmoplast, a disclike membrane-bound structure containing polysaccharides and glycoproteins.

E. results in increasingly smaller daughter cells with each successive round of the cell cycle.

MEIOSIS (Pages 563–567)
Homologous Chromosomes Pair Off During Meiosis (Pages 563–564)

17–20   Easy, multiple choice

Two homologous chromosomes:

A. are identical in many regions.

B. pair during mitosis.

C. form bivalents in meiosis.

D. are both present in a haploid cell.

E. are able to recognize each other because the DNA sequences at their centromeres are identical.

17–21   Intermediate, short answer

Fill in the blanks in the following sentences.

"An organism that has five homologous pairs of chromosomes can produce at least _____ genetically different types of gametes. Recombination will _____ the number of genetically different gametes possible."

Meiosis Involves Two Cell Divisions Rather Than One (Pages 564–567)

17–22   Difficult, multiple choice + short answer

In some fungi, orderly cell division during meiosis gives rise to a row of four haploid spores in each spore sac as shown in Figure Q17–22A. You notice that a strain of the fungus produced by crossing a brown strain with white strain gives rise mostly to spore sacs as shown in Figure Q17–22B, with a few spore sacs like those in Figure Q17–22C. Which of the following can be concluded from the above observations? Explain your reasoning.

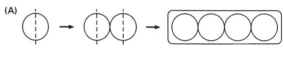

A. Meiosis I and meiosis II in your fungus occur in the reverse order from that which occurs in humans.

B. Recombination in the fungus can occur during prophase I.

C. Recombination in the fungus cannot occur during prophase II.

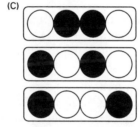

D. Recombination has occurred between the centromere and the gene responsible for spore color.

E. Recombination has occurred between the gene responsible for spore color and the end of the chromosome arm.

Q17–22

17–23   Intermediate, multiple choice + short answer

In mammals, there are two sex chromosomes, X and Y, which behave like homologous chromosomes during meiosis. Normal males have one X chromosome and one Y chromosome, and normal females have two X chromosomes. Males with an extra Y chromosome (XYY) occasionally are found. Which of the following could give rise to such an XYY male? Explain your answer.

A. Nondisjunction in the first meiotic division of spermatogenesis; normal meiosis in the mother.

B. Nondisjunction in the second meiotic division of spermatogenesis; normal meiosis in the mother.

C. Nondisjunction in the first meiotic division of oogenesis; normal meiosis in the father.

D. Nondisjunction in the second meiotic division of oogenesis; normal meiosis in the father.

E. Nondisjunction in the first meiotic division of both the mother and father's gametes.

Answers

A17–1.   In the eucaryotic cell cycle, DNA replication occurs during <u>S phase</u>, nuclear and cytoplasmic division occur during <u>M phase</u>, and cell growth occurs during <u>interphase</u>. In a typical somatic cell, <u>G_1 phase (interphase is also acceptable)</u> is the longest phase of the cell cycle.

A17–2.   The cells produced would get smaller and smaller.

A17–3.   A and E. At 37°C, the cells all have one genome worth of DNA, meaning that they have not replicated their DNA and therefore have not entered S phase. Hence they could either have a defect in initiation of DNA replication (A) or a defect that causes inappropriate delay in G_1 (E).

A17–4.   E. The contractile ring is found only in animal cells. The mitotic spindle is composed of microtubules (not actin) (A), assembles in early mitosis (B), and is not formed in procaryotes (which have no microtubules) (C). The contractile ring is made up of actin and myosin (D).

A17–5.   B and G. The ER and the nuclear membrane (which is continuous with the ER) both fragment at mitosis. Mitochondria, lysosomes, ribosomes, and chloroplasts are all small and numerous organelles and do not need to be broken into small pieces to ensure roughly equal segregation to both daughter cells. Centrosomes do not fragment randomly: they duplicate themselves in an orderly and elaborate way that we do not understand.

A17–6.   Prophase, prometaphase, metaphase, anaphase, telophase, cytokinesis.

A17–7.   A = telophase; B = prophase; C = anaphase; D = prometaphase.

A17–8.   A, B, and D. The centrosome duplicates (B), and some microtubules interact with each other to form the mitotic spindle (D). The number of microtubules per centrosome increases during prophase and microtubules on average are shorter during prophase than during interphase.

A17–9.   E. Kinetochore microtubules cannot form if the chromosomes are separated from the microtubules in the cytoplasm by the nuclear envelope. The nuclear envelope disassembles during prometaphase (so A is untrue) by breaking up into vesicles containing lipids from both the outer and inner envelope (so B is untrue). Integral membrane proteins of the nuclear envelope and some of the nuclear lamins remain associated with the vesicles (C). Phosphorylation of lamins (not integrins) triggers breakdown of the nuclear lamina (D).

A17–10. (A) 16. After DNA replication there will be 16 chromatids in this cell, and each chromatid has one kinetochore. (B) 8. After the first meiotic division, each daughter cell contains one set of four duplicated chromosomes, that is, eight chromatids. These enter the second meiotic division without further DNA replication.

A17–11. A. The DNA-binding proteins of the kinetochore recognize the centromere, not the telomere (B), and thus cannot bind to chromosomes lacking centromeres (D). If kinetochore proteins bound to the very tips of microtubules, they would soon fall off, since the microtubules are dynamic and are constantly losing the monomers at the ends of the filament (C). Kinetochores bind to free microtubules, converting them to kinetochore microtubules (E).

A17–12. C. When the attachment to one kinetochore is severed the whole chromosome moves to the opposite pole, showing that the kinetochore microtubules are pulling on their attached chromatid, not pushing it. The jiggling movement (A) is simply a sign that the chromosomes are subject to forces from both sides. When the attachment between sister chromatids is severed (B), both daughter chromosomes move toward their respective poles, but they could either be pushed there or pulled there. Similarly the shape of the chromosomes as they move toward the pole (D) indicates that the chromosomes are being moved via an attachment at the centromere, but does not indicate whether they are being pushed or pulled. The disappearance of the spindle when colchicine is added to the cells simply shows that the spindle is composed of microtubules (E).

A17–13.   E. Anaphase is triggered by proteases that cleave the proteins that hold the sister chromatids together. Anaphase A and anaphase B generally occur at the same time. In cells where anaphase B predominates, the spindle will elongate more than in cells where anaphase A predominates. In anaphase A, only the kinetochore microtubules shorten.

A17–14.   D. The motor protein used in anaphase B that pushes the polar microtubules apart moves toward the plus end of microtubules, whereas motor proteins used in anaphase A move toward the minus end. Myosin (A) is not involved in either anaphase A or anaphase B. Motor proteins require ATP hydrolysis (are ATPases) and are used in both anaphase A and anaphase B, so both types of anaphase will be affected by ATPγS (B). The motor protein used in anaphase A moves toward the minus end of microtubules as does the motor protein attached to the cell cortex used in anaphase B (C). Microtubules are also used in both anaphase A and anaphase B, so colchicine will affect both (E).

A17–15.   A and D. The chromosomes do not decondense and the lamina cannot be assembled until the nucleus is complete and nuclear proteins (including most of the lamins) have been imported through the pores. Transcription does not begin until the chromosomes decondense. Both the contractile ring in animals and the phragmoplast in plants begin to assemble before the nuclear envelope has re-formed.

A17–16.   B. A Ser to Ala mutation in the phosphorylation site of a lamin will prevent it from being phosphorylated. Since the nonphosphorylated form is the form that polymerizes into a lattice, such a mutation will stabilize the nuclear lamina and prevent its dissasembly and nuclear envelope breakdown at prometaphase. Telophase involves repolymerization of lamins and therefore will not be blocked by the mutation (A). Lamins are not part of the nuclear pore complex (C). The mitotic spindle begins to form before the nucleus has broken down, forming a sort of cage around the nucleus (D). Lamins are not involved in chromosome condensation (E).

A17–17.   A, C, and D. The cleavage furrow begins to form before the end of telophase. Although the furrow eventually becomes independent of the mitotic spindle, the furrow requires the mitotic spindle for its establishment. The furrow always forms perpendicular to the polar microtubules about midway between the poles, but since the spindle is displaced in some cells, cell division does not always occur down the middle of the cell.

A17–18.   A and D. All cell movement requires ATP, and in cytokinesis, actin and myosin molecules are moving relative to one another to cause contraction of the contractile ring. The assembly of the contractile ring requires a general rearrangement of the filaments in the cell cortex. The contractile ring completely disassembles after mitosis (B). Phosphorylation of integrins, which weakens their hold on the extracellular matrix and allows cells to round up, generally *precedes* cytokinesis and is part of the general rearrangement of cell structure (including the cell cortex) that accompanies cell division (C). Microtubules do not play an important role in animal cytokinesis (E).

A17–19.   A and C. Plant cells do not use a contractile ring to divide their cells. However, division still takes place perpendicular to the long axis of the mitotic spindle because the phragmoplast—the remnants of the polar microtubules (D)—direct the Golgi vesicles that form the new cell wall and membrane to construct a plate perpendicular to the spindle. Movement of the vesicles, of course, involves motor proteins and thus requires ATP (B). The presence of a cell wall, although restricting the orientation of plant cells with respect to one another, does not prevent cell growth and expansion between divisions (E).

A17–20. A and C. Homologous chromosomes do not pair in mitosis. Only one chromosome of a pair is present in a haploid cell. It is unknown how homologous chromosomes recognize each other.

A17–21. <u>32</u>; <u>increase</u>. Since homologous chromosomes assort randomly at meiosis, and a gamete has two choices for each chromosome (since sexual organisms are diploid), there are 2^5, or 32, possible genetically different gametes. Recombination effectively increases the number of types of chromosomes and therefore increases the number of possible genetically different gametes.

A17–22. B and D. In normal meiosis, the chromosomes are distributed as shown in Figure A17–22A to give the pattern of spores that was shown in the question (Figure Q17–22B). (The dark-colored chromosome carries the allele for dark pigment.) This is what is observed in your fungus most of the time.

In order to get mostly normal but just a few oddball spore sacs, there must be recombination between the two homologous chromosomes. This recombination must occur when both homologous chromosomes are in the same cell (i.e., in prophase I of meiosis). D is correct and E is incorrect, since, if recombination had taken place only between the gene for spore color and the end of that chromosome arm, it would look as if no recombination had taken place (Figure A17–22B).

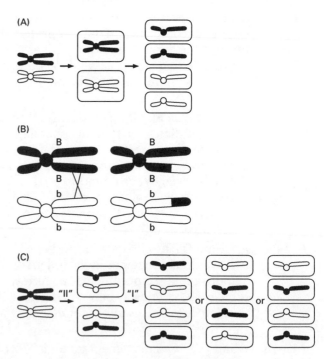

If meiosis I and meiosis II were reversed (A), we would expect a completely random pattern of spores in spore sacs, since the two homologues could go to either cell in the second division, as shown in Figure A17–22C.

During prophase II, there is only one homologue in each cell; if recombination took place between the two sister chromatids of this homologue we would not be able to detect it, as both homologues are identical. Thus we cannot absolutely say that recombination has not taken place in prophase II (C).

A17–22

A17–23. B. Nondisjunction in one or the other meiotic division will give rise to gametes with two sex chromosomes instead of one. Since the only source of Y chromosomes is the father, the gametes that produce an XYY male must be X (a normal egg) and YY. Nondisjunction in the first meiotic division would result in two XY sperm and two sperm with no sex chromosome. Nondisjunction in the second meiotic division, in contrast, could give rise to YY gametes:

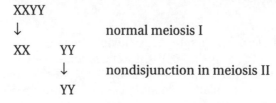

XXYY
↓   normal meiosis I
XX   YY
↓   nondisjunction in meiosis II
YY

18 Cell-Cycle Control and Cell Death

Questions

THE CELL-CYCLE CONTROL SYSTEM (Pages 572–581)
A Central Control System Triggers the Major Processes of the Cell Cycle (Pages 572–574)

18–1 Easy, matching/fill in blanks

For each of the following sentences, fill in the blanks with the correct word or phrase.

    A. The four phases of the cell cycle, in order, are G_1, _____, _____, and _____.

    B. A cell contains the most DNA after _____ phase of the cell cycle.

    C. A cell is smallest in size after _____ phase of the cell cycle.

    D. Growth occurs in _____, _____, and _____ phases of the cell cycle.

    E. A cell does not enter mitosis until it has completed _____ synthesis.

18–2 Difficult, short answer

Diploid cells from *Drosophila* normally have 8 chromosomes (2 haploid sets). *Drosophila* cell lines can be produced that have 16 chromosomes (4 haploid sets). The latter cells grow to about twice the size of normal *Drosophila* cells. From this observation, suggest a mechanism for how the cell might determine its own size at a G_1 cell-size checkpoint.

18–3 Intermediate, short answer (Requires information from Chapter 17)

What would happen to the progeny of a cell that proceeded to mitosis and cell division after entering S phase but not completing it? Keep in mind that highly condensed chromatin, including the centromere region, is replicated late in S phase. Explain your answer.

The Cell-Cycle Control System Is Based on Cyclically Activated Protein Kinases (Pages 574–575)

18–4 Easy, multiple choice

Levels of cyclin-dependent kinase (Cdk) activity change during the cell cycle as a result of:

    A. the Cdks phosphorylating each other.

    B. the Cdks activating the cyclins.

    C. the level of production of Cdks changing throughout the cell cycle.

    D. the Cdks binding different cyclins to become active.

    E. changing levels of cyclin phosphorylation.

MPF Is the Cyclin-Cdk Complex That Controls Entry into M Phase (Pages 575–576)

18–5   Easy, multiple choice

M-phase promoting factor:

    A. is a kinase.

    B. contains a kinase.

    C. was purified from extracts of frog oocytes in G_2 phase.

    D. is found only in *Xenopus*.

    E. acts to initiate both S phase and M phase.

18–6   Easy, short answer

Name three effects of MPF activity that are required for cells to enter mitosis.

Cyclin-dependent Protein Kinases Are Regulated by the Accumulation and Destruction of Cyclin (Pages 576–577)

18–7   Easy, multiple choice

The concentration of mitotic cyclin:

    A. rises dramatically during M phase.

    B. changes as Cdk concentration changes.

    C. falls during M phase due to ubiquitination.

    D. is highest in G_1 phase.

    E. determines the timing of mitosis in the cell cycle.

The Activity of Cdks Is Further Regulated by Their Phosphorylation and Dephosphorylation (Page 578)

18–8   Intermediate, multiple choice

You have isolated a strain of mutant yeast that divides normally at 30°C but arrests in the cell cycle after S phase at 37°C. You have isolated its mitotic cyclin and mitotic Cdk and find that both are perfectly normal and form a normal MPF complex at both temperatures. Which of the following temperature-sensitive mutations could be responsible for the behavior of this strain of yeast?

    A. Inactivation of an enzyme that ubiquinates the MPF cyclin.

    B. Inactivation of a protein kinase that acts on the MPF Cdk kinase.

    C. Inactivation of a protein kinase that phosphorylates the MPF cyclin.

    D. Inactivation of a phosphatase that acts on the MPF Cdk kinase.

    E. The continuous activation of a phosphatase that removes all phosphate groups from MPF.

Different Cyclin-Cdk Complexes Trigger Different Steps in the Cell Cycle (Pages 578–579)

18–9   Easy, art labeling

Match the following labels to the numbered label lines on Figure Q18–9:

A. G_1 phase.

B. Mitotic cyclin.

C. Mitotic Cdk.

D. S phase.

E. S phase cyclin.

F. S phase Cdk.

G. MPF.

H. G_2 phase.

Q18–9

18–10   Intermediate, data interpretation (Requires information from section on pages 576–577)

You have cloned a cyclin gene from a mouse and want to find out where in the cell cycle the cyclin acts. To do this you introduce the cloned gene into mutant yeast cells, each of which is mutant in one of the cyclins. The experimental data are given in Table Q18–10. Where in the mouse cell cycle does your cyclin act?

Table Q18–10

Yeast mutant cells	Phenotype of yeast cells without mouse cyclin gene	Phenotype of yeast with introduced mouse cyclin gene
1.	Cells arrested in G_1 phase	Cells arrested in G_1 phase
2.	Cell arrested in G_2 phase	Cells arrested in G_2 phase
3.	Cells arrested in G_1 phase and smaller than normal	Cells grow to normal size and progress through the cell cycle normally

The Cell Cycle Can Be Halted in G_1 by Cdk Inhibitor Proteins (Pages 580–581)

18–11   Easy, short answer

Place the following events in the order in which they occur when cells stop dividing at a G_1 checkpoint after the cells have been irradiated.

A. Production of p21.

B. DNA damage.

C. Inactivation of cyclin-Cdk complex.

D. Activation of p53.

Cells Can Dismantle Their Control System and Withdraw from the Cell Cycle (Page 581)

18–12   Easy, multiple choice

Cells in the G_0 state:

    A. cannot reenter the cell cycle.

    B. can remain in that state for a lifetime.

    C. have entered from G_1 or G_2 checkpoints.

    D. do not divide.

    E. have duplicated their DNA.

18–13   Intermediate, short answer

What is the main molecular difference between cells in a G_0 state and cells that have simply paused in G_1?

CONTROL OF CELL NUMBERS IN MULTICELLULAR ORGANISMS (Pages 582–589)

18–14   Easy, short answer

The number of cells in an adult tissue or animal is determined by regulated cell proliferation. What else helps determine cell numbers?

Cell Proliferation Depends on Signals from Other Cells (Pages 582–584)

18–15   Easy, multiple choice

In proliferating cells:

    A. Rb protein is active.

    B. nutrients are limiting.

    C. growth factors secreted by other cells are bound to receptors.

    D. cyclin-Cdk complexes are inactive.

    E. all Cdks are continuously active.

18–16   Intermediate, multiple choice + short answer

Which of the following mutations would be likely to lead to a cell that became arrested in the cell cycle? For each of these cases say at which of the checkpoints (early G_1, late G_1, or G_2) the cell would become arrested.

    A. A mutation in a cell-surface growth factor receptor that made it active in the absence of growth factor.

    B. A mutation that destroyed the kinase activity of the S-phase cyclin-Cdk complex.

C. A mutation that made the kinase of the G_1 cyclin-Cdk complex continuously active.

D. A loss of the phosphorylation sites on the Rb protein.

E. A defect in the Rb protein that made it unable to bind to gene regulatory proteins.

Animal Cells Have a Built-in Limitation on the Number of Times They Will Divide (Page 584)

18–17 Easy, short answer

Which of the following statements are true?

A. Cells from mice and humans will undergo the same maximum number of cell divisions in culture.

B. Cells taken from the same animal at different ages will undergo the same maximum number of cell divisions in culture.

C. Normal cells stop proliferating after a limited number of divisions.

D. When cells become senescent they can no longer divide in response to growth factors.

E. Cells become senescent once they have used up the growth factors in the medium.

18–18 Intermediate, multiple choice + short answer (Requires information from sections on pages 584–585 and pages 587–589)

If cells isolated from an animal become senescent after a limited number of divisions in culture, how are the cell lines used in experimental cell biology able to proliferate indefinitely in culture? Explain your answer.

A. They are continually given fresh supplies of growth factor.

B. They have undergone mutations that inactivate the genes involved in programmed cell death.

C. They have undergone mutations that abolish cell senescence.

D. They have become differentiated.

E. They accumulate Cdk inhibitor proteins.

Animal Cells Require Signals from Other Cells to Avoid Programmed Cell Death (Pages 584–585)

18–19 Easy, multiple choice

Programmed cell death occurs:

A. rarely and selectively where survival factors are not present.

B. to eliminate unneeded cells.

C. only in unhealthy or abnormal cells.

D. only during embryonic development.

E. by means of an intracellular suicide program.

18–20   Easy/intermediate, short answer

What is the cause of the massive amount of programmed cell death of nerve cells (neurons) that occurs in the developing vertebrate nervous system, and what purpose does it serve?

Programmed Cell Death Is Mediated by an Intracellular Proteolytic Cascade (Page 585–587)

18–21   Easy, multiple choice (Requires information from section on pages 584–585)

Apoptosis differs from necrosis in that:

    A. necrosis happens more frequently.

    B. necrosis causes DNA to fragment.

    C. in necrosis cells swell and burst, while in apoptosis cells shrink and condense.

    D. necrosis uses a kinase cascade and apoptosis uses a protease cascade.

    E. necrosis contributes to cancer when deregulated.

    F. necrosis occurs only in injured or abnormal cells.

18–22   Easy, multiple choice

Activated intracellular proteases directly mediate programmed cell death by:

    A. cleaving other proteases.

    B. digesting the nuclear DNA.

    C. activating phagocytosis by a neighboring cell.

    D. destroying survival signals.

    E. eliciting killer signals from other cells.

    F. cleaving various key proteins in the cell.

Cancer Cells Disobey the Social Controls on Cell Proliferation and Survival (Pages 587–589)

18–23   Intermediate, short answer

The protein p53 is activated when DNA is damaged and helps arrest the cell cycle in G_1, allowing time for the cell to repair its DNA before replicating it. Activated p53 arrests the cell cycle by stimulating the transcription of the gene that encodes the Cdk inhibitor protein p21. Mutations that inactivate p53 contribute to 50% of human cancers. Would you classify p53 as a tumor-suppressor gene or a proto-oncogene? Explain your answer.

Answers

A18–1. A. The four phases of the cell cycle, in order, are G_1, S phase, G_2, and M phase.

B. A cell contains the most DNA after S phase of the cell cycle.

C. A cell is smallest in size after M phase of the cell cycle.

D. Growth occurs in G_1, S, and G_2 phases of the cell cycle.

E. A cell does not enter mitosis until it has completed DNA synthesis.

A18–2. The amount of DNA remains constant during G_1 whereas the amounts of most proteins steadily increase as the cell grows. The observation that the size of the *Drosophila* cells is roughly proportional to their total amount of DNA raises the possibility that the size control mechanism might depend on somehow comparing the amount of a protein to the amount of DNA. A protein X that is required for driving cells into S phase, for example, may be inactivated by binding to sites on DNA, so that only when the level of X rises high enough that all the binding sites on DNA are occupied can free X initiate S phase. In cells that have twice the normal amount of DNA, there will be twice as many X-binding sites, so that the cell will have to grow twice as large with twice the normal amount of X before S is triggered. The actual mechanism of the G_1 cell-size checkpoint is not known.

A18–3. The daughter cells would probably die. Those chromosomes that had not completed replication in S phase would have only one centromere, as the centromere is the last part of the chromosome to be replicated, and would therefore be segregated to only one of the two daughter cells at random. At least one, and probably both, of the daughter cells would thus receive an incomplete set of chromosomes and would be unlikely to be viable. Even if one daughter cell, by chance, received a full set of chromosomes, some of these chromosomes would be incompletely replicated and the cell would probably still not be viable.

A18–4. D.

A18–5. B.

A18–6. Phosphorylation of lamins, which leads to nuclear envelope breakdown; phosphorylation of chromosomal proteins, causing the chromosomes to condense; phosphorylation of microtubule-associated proteins, which facilitates the formation of the mitotic spindle.

A18–7. C and E.

A18–8. B, D, and E. The Cdk in the MPF complex has to be phosphorylated at some sites and dephosphorylated at others in order to be active. Thus mutations that inactivated either protein kinase responsible for the activating phosphorylation (B) or the protein phosphatase responsible for the activating dephosphorylation (D) could give this defect. A mutation that led to the continuous activation of a phosphatase that removes all of the phosphate from the Cdk (E) could also explain the defect. A cell with a mutation like that in (A) would enter mitosis. Cyclin does not have to be phosphorylated to be active so (C) would have no effect.

A18–9. A, 5; B, 6; C, 3; D, 2; E, 4; F, 1; G, 7; H, 8.

A18–10. At the first G_1 checkpoint, where the cell checks that it is sufficiently large to continue through G_1.

A18–11. B, D, A, and C.

A18–12. B and D.

A18–13. In G_0 the cell-cycle control system is partly dismantled, in that some of the Cdks and cyclins are not present. Cells paused in G_1, by contrast, still contain all the components of the cell-cycle control system. Whereas the latter cells can rapidly start to progress through the cycle when conditions are right, G_0 cells need to synthesize the missing cell-cycle control proteins in order to reenter the cycle, which usually takes 8 hours or more.

A18–14. Programmed cell death also helps to adjust cell numbers. The cells require signals from other cells to avoid programmed cell death so that the levels of such survival signals helps determine how many cells live and how many die.

A18–15. C.

A18–16. B, late G_1. D, early G_1.

A18–17. C and D. E is untrue; although cells will stop dividing if the supply of growth factors is exhausted, if they have not yet reached their maximum number of divisions, they will resume dividing when more growth factor is supplied.

A18–18. C. Cell lines that can grow indefinitely in culture have become "immortalized" by mutations in one or more proliferation or antiproliferation genes. Senescent cells become insensitive to growth factor, so that once the maximum number of cell divisions is reached providing more growth factor has no effect (A). Senescence is not the same phenomenon as programmed cell death, so mutations affecting the ability to undergo programmed cell death would not have an effect on senescence (B).

A18–19. B and E.

A18–20. Immature neurons are produced in excess of the number that will eventually be required. They compete for the limited amount of survival factors secreted by the target cells they contact. Those cells that fail to get enough survival factor undergo programmed cell death. Up to half or more of the nerve cells generated die in this way. This competitive mechanism helps match the number of developing nerve cells to the number of target cells they contact.

A18–21. C and F.

A18–22. A and F.

A18–23. Because one of the normal functions of the p53 protein is to suppress cell division and its inactivation promotes the development of cancer, it is classified as a tumor-suppressor gene. For a proto-oncogene to contribute to cancer development it has to be activated by mutation to become an oncogene. The normal function of a proto-oncogene is often to promote cell division, not to inhibit it.

19 Tissues

Questions

EXTRACELLULAR MATRIX AND CONNECTIVE TISSUES (Pages 594–605)
Plant Cells Have Tough External Walls (Pages 594–596)

19–1   Easy, multiple choice

Both multicellular plants and animals:

    A. have cells capable of locomotion.

    B. have cells with cell walls.

    C. have vascular systems containing circulating cells.

    D. have a cytoskeleton composed of actin filaments, microtubules and intermediate filaments.

    E. have tissues composed of multiple different cell types.

19–2   Easy, matching/fill in blanks

For each of the following sentences, fill in the blanks with the correct word or phrase selected from the list below. Use each word or phrase once only.

    A. Except at a meristem, most of the growth in the size of a plant tissue is the result of cell _____ rather than cell _____.

    B. The swelling pressure that drives plant cell growth is generated by _____.

    C. The _____ plant cell wall is expandable to allow growth of the cell.

    D. A plant wilts as a result of the loss of _____ in its cells.

    E. The cell walls of xylem cells are thickened and strengthened by the deposition of _____.

division; tertiary; osmosis; wax; primary; enlargement; progression through the cell cycle; secondary; turgor pressure; DNA synthesis; lignin.

Cellulose Fibers Give the Plant Cell Wall Its Tensile Strength (Pages 596–600)

19–3   Intermediate, multiple choice (Requires information from Chapter 14)

The cellulose fibers in plant cell walls are synthesized:

    A. in the Golgi apparatus and transported to the cell surface in membrane vesicles.

    B. in the cytosol and transported to the cell surface via microtubules.

    C. by enzymes transported to the plasma membrane via the Golgi apparatus.

    D. by cytosolic enzymes that gather at the plasma membrane.

    E. in the endoplasmic reticulum and transported to the plasma membrane.

19–4   Easy, short answer

(A) Indicate the direction in which the plant cell shown in Figure
Q19–4 is most likely to grow. (B) Explain your answer.

Q19–4

19–5   Intermediate, short answer

Why would altering the orientation of microtubules on the underside of the plant cell plasma
membrane affect the direction in which the plant cell would grow?

Animal Connective Tissues Consist Largely of Extracellular Matrix (Page 600)

19–6   Easy, multiple choice

A major distinction between the connective tissues in an animal and other main
tissue types such as epithelia, nervous tissue, or muscle is:

A. the ability of connective tissue cells such as fibroblasts to change shape.

B. the abundant extracellular matrix in connective tissues.

C. that connective tissues can withstand mechanical stresses.

D. the exceptional amounts of polysaccharides in the extracellular matrix of connective tissues.

E. the numerous connections connective tissue cells make with each other.

Collagen Provides Tensile Strength in Animal Connective Tissues (Pages 600–602)

19–7   Easy, short answer (Requires information from Chapter 16)

What are the main structures providing tensile strength in:

(A) animal connective tissue?

(B) animal epithelium?

(C) plant cell walls?

19–8   Easy, short answer

In what form is collagen secreted from fibroblasts, and what happens to it after secretion to
allow formation of large fibers?

Cells Organize the Collagen That They Secrete (Pages 602–603)

19–9   Easy, multiple choice

Fibroblasts organize the collagen of the extracellular matrix by:

- A. cutting and rejoining the fibrils.
- B. crawling along existing fibrils and depositing new collagen as they go.
- C. twisting fibrils together to make ropelike fibers.
- D. pulling the collagen into sheets or cables after it has been secreted.
- E. controlling the local Ca^{2+} concentration so as to regulate collagen assembly.

Integrins Couple the Matrix Outside a Cell to the Cytoskeleton Inside It (Pages 603–604)

19–10   Easy, multiple choice

Cells bind to the extracellular matrix via:

- A. fibronectin in the plasma membrane.
- B. integrins in the plasma membrane.
- C. phospholipids in the plasma membrane.
- D. actin filaments.
- E. microtubules.

Gels of Polysaccharide and Protein Fill Spaces and Resist Compression (Pages 604–605)

19–11   Easy, multiple choice

Proteoglycans in the extracellular matrix of animal tissues:

- A. provide tensile strength.
- B. form a dense compact matrix.
- C. are linked to microtubules through the plasma membrane.
- D. are polysaccharides composed of glucose subunits.
- E. provide cartilage with the ability to resist compression.

19–12   Intermediate, multiple choice

A typical proteoglycan molecule:

- A. is a long unbranched polymer of glycosylated amino acids.
- B. binds strongly to cell surfaces because it carries a large positive charge.
- C. can occupy a volume as large as 10^9 μm^3.
- D. consists chiefly of protein.
- E. is accompanied by a cloud of positively charged ions.

19–13   Intermediate, short answer

Why are proteoglycans excellent space-fillers in the extracellular matrix?

EPITHELIAL SHEETS AND CELL-CELL JUNCTIONS (Pages 605–613)
Epithelial Sheets Are Polarized and Rest on a Basal Lamina (Pages 606–607)

19–14   Easy, short answer

Give three functions of epithelia in animals.

19–15   Easy, multiple choice

A basal lamina:

    A. is a thin layer of connective tissue cells and matrix underlying an epithelium.
    B. is a thin layer of extracellular matrix underlying an epithelium.
    C. is attached to the apical surface of an epithelium.
    D. is impermeable to small organic molecules.
    E. separates epithelial cells from each other.

Tight Junctions Make an Epithelium Leak-proof and Separate Its Apical and Basal Surfaces (Pages 607–609)

19–16   Easy, art labeling + data interpretation

Label the five different types of cell-cell junctions shown in Figure Q19–16, and identify the apical and basal surfaces of the epithelium.

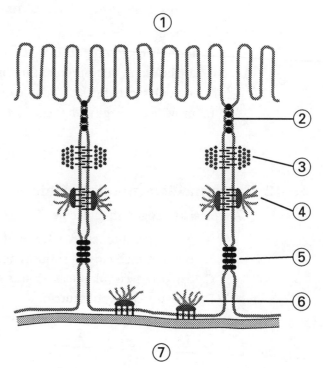

Q19–16

19–17   Easy, multiple choice

Tight junctions between epithelial cells serve to:

    A. transport small molecules from the apical to the basal side of the epithelium.

    B. bind the cells strongly together so that they cannot be torn apart.

    C. seal cells together so that water-soluble molecules will not leak across the epithelium between the cells.

    D. allow small molecules and ions to pass from one epithelial cell to its neighbor.

    E. keep ions from diffusing out from the cell into the extracellular medium.

19–18   Easy, short answer

How would you demonstrate experimentally the existence of functional tight junctions between neighboring cells in an epithelium?

Cytoskeleton-linked Junctions Bind Epithelial Cells Robustly to One Another and to the Basal Lamina (Pages 609–612)

19–19   Easy, multiple choice

Adherens junctions:

    A. connect epithelial cells to the basal lamina.

    B. join adjacent epithelial cells to each other via cadherins.

    C. bind intermediate filaments on the intracellular side of the membrane.

    D. are most similar to tight junctions.

    E. connect epithelial cells to each other through integrins.

19–20   Intermediate, short answer

(A) Adherens junctions and desmosome junctions have a similar function in epithelia. What is it?

(B) What features do these two junctions have in common that enable them to serve this function?

(C) Which of the other cell junctions serves a similar function?

Gap Junctions Allow Ions and Small Molecules to Pass from Cell to Cell (Pages 612–613)

19–21   Intermediate, multiple choice

Which of the following do gap junctions allow to pass?

    A. Most enzymes.

    B. Plasma membrane proteins.

    C. Cyclic AMP.

    D. Hydrophobic molecules of less than 1000 daltons.

    E. RNA.

19–22 Intermediate, matching/fill in blanks (Requires information from sections on pages 596–612)

Match the molecules on the left with the cell structures in which they reside (on the right), by writing the appropriate number beside each molecule. Note: you can use a structure more than once, and more than one structure can contain a given molecule.

A.	Cadherin.	1.	Extracellular matrix.
B.	Cellulose.	2.	Cytoskeleton.
C.	Collagen.	3.	Cell wall.
D.	Connexon protein.	4.	Desmosome junction.
E.	Fibronectin.	5.	Basal lamina.
F.	Integrin.	6.	Adherens junction.
G.	Keratin.	7.	Gap junction.
H.	Laminin.	8.	Hemidesmosome junction.
I.	Lignin.		

TISSUE MAINTENANCE AND RENEWAL, AND ITS DISRUPTION BY CANCER (Pages 613–621)
Different Tissues Are Renewed at Different Rates (Page 615)

19–23 Easy, short answer

Place the following in order of their replacement times from shortest to longest:

    A. Epidermal cell.

    B. Nerve cell.

    C. Bone matrix.

    D. Red blood cell.

    E. Cell lining the gut.

Stem Cells Generate a Continuous Supply of Terminally Differentiated Cells (Pages 615–618)

19–24 Easy, multiple choice

Stem cells:

    A. cycle between an undifferentiated and a differentiated state.

    B. are required to renew all cell types.

    C. look like the specialized cells of the tissue.

    D. migrate to where they are needed.

    E. can sometimes have progeny of more than one cell type.

Mutations in a Single Dividing Cell Can Cause It and Its Progeny to Violate Normal Controls (Pages 618–619)

19–25   Easy, multiple choice

A malignant tumor is more dangerous than a benign tumor because:

    A. its cells are proliferating faster.

    B. it metastasizes by traveling through the digestive tract.

    C. it causes neighboring cells to mutate.

    D. it causes neighboring normal tissue cells to die.

    E. its cells invade other tissues.

19–26   Intermediate, short answer (Requires information from section on pages 603–604)

A certain mutation in the receptor for epidermal growth factor (EGF) leads the mutated receptor protein to send a positive signal to its associated intracellular signaling pathway even when EGF is not bound to it. This leads to abnormal cell proliferation in the absence of growth factor. On the basis of this information, would you class the gene for the EGF receptor as a tumor-suppressor gene or a potential oncogene? Explain your answer.

Cancer Requires an Accumulation of Mutations (Pages 620–621)

19–27   Easy, multiple choice

A normal cell is converted into a cancer cell by:

    A. a loss-of-function mutation in an oncogene.

    B. any mutation that allows the cell to divide more frequently than normal.

    C. an accumulation of mutations affecting several different genes.

    D. loss of one of the two copies of a tumor suppressor gene.

    E. the simultaneous triggering of multiple mutations by environmental mutagens such as tobacco and radioactivity.

19–28   Intermediate, multiple choice

Two people who both have an inherited predisposition to colorectal cancer, resulting from the same mutation in the APC gene, get married. Their children:

    A. will all have an inherited predisposition to colorectal cancer.

    B. will not survive to birth.

    C. have a 25% chance of being normal.

    D. if affected at all, will show the disease in a more severe form than their parents.

    E. have a 75% chance of being normal.

DEVELOPMENT (Pages 621–628)
Programmed Cell Movements Create the Animal Body Plan (Page 622)

19–29   Easy, multiple choice (Requires information from section on pages 609–612)

Gastrulation is:

A. a process peculiar to fish.

B. the first stage of embryonic development following fertilization.

C. the process by which cells of the embryo first become organized into an epithelial sheet.

D. a coordinated movement of the cells of the early embryo, through which the rudiment of the gut is formed.

E. a prelude to cleavage.

19–30   Intermediate, short answer (Requires information from Chapter 17)

In the cleavage stage of embryonic development of many animals, the fertilized egg divides rapidly into a ball of small cells that is initially no bigger than the fertilized egg itself. Describe the cell cycles that are producing these cells; what is unusual about them?

Cells Switch On Different Sets of Genes According to Their Position and Their History (Pages 622–624)

19–31   Easy, multiple choice

During pattern formation, cells become different from each other by:

A. secreting different amounts of signaling molecules.

B. activating or repressing different sets of genes.

C. asymmetric divisions in which the sister cells inherit different sets of genes.

D. controlling their adhesiveness to other cells.

E. controlled mutations in their DNA.

Diffusible Signals Can Provide Cells with Positional Information (Pages 624–626)

19–32   Easy, multiple choice

The secreted signaling protein Sonic hedgehog provides embryonic cells with positional information by:

A. binding to DNA and controlling gene expression.

B. diffusing to form an extracellular concentration gradient.

C. regulating cell movements

D. triggering nerve signals in the embryonic nervous system.

E. conferring the character of floorplate on any cell that is exposed to a high concentration of Sonic hedgehog, and the character of motor neuron on any cell that is exposed to a low concentration.

Studies in *Drosophila* Have Given a Key to Vertebrate Development (Pages 626–627)

19–33   Intermediate, short answer (Requires information from Chapter 10)

You have isolated the DNA for a developmental gene from *Drosophila* and have discovered that a homologous gene is present in the mouse genome.

(A) What technique would you use to find where the mouse gene was expressed in the mouse embryo?

(B) Having found that the gene was expressed at several different sites in the mouse embryo, how might you then proceed to test what functions the gene had in mouse development?

Similar Genes Are Used Throughout the Animal Kingdom to Give Cells an Internal Record of Their Position (Pages 627–628)

19–34   Intermediate, short answer

Why does a mutation in one of the Hox genes in *Drosophila* produce a fly with legs on its head instead of antennae?

Answers

A19–1.   E. D is incorrect, as plant cells do not have intermediate filaments.

A19–2.   A. Except at a meristem, most of the growth in the size of a plant tissue is the result of cell <u>enlargement</u> rather than cell <u>division</u>.

B. The swelling pressure that drives plant cell growth is generated by <u>osmosis</u>.

C. The <u>primary</u> plant cell wall is expandable to allow growth of the cell.

D. A plant wilts as a result of the loss of <u>turgor pressure</u> in its cells.

E. The cell walls of xylem cells are thickened and strengthened by the deposition of <u>lignin</u>.

A19–3.   C. Cellulose fibers are synthesized at the cell surface by protein complexes embedded in the plasma membrane. All proteins embedded in the plasma membrane must have been transported there via the Golgi apparatus.

A19–4.   (A) Figure A19–4. (B) Cellulose fibers are highly resistant to stretching and thus a plant cell tends to grow, under the stimulus of turgor pressure, in a direction perpendicular to the orientation of the fibers in the cell wall.

A19–4

A19–5.   A plant cell tends to enlarge in a direction perpendicular to the orientation of cellulose fibers in the cell wall. The orientation of the cellulose fibers is determined by the direction of movement in the plasma membrane of the complexes of cellulose synthase that synthesize cellulose. This in turn is guided by the orientation of microtubules underlying the membrane in such a way that cellulose fibrils become laid down parallel to the microtubules.

A19–6.   B. Cells in many other types of tissue can change shape (e.g., muscle cells when they contract), and can withstand mechanical stress (e.g., cells of the epidermis). Connective tissue cells tend to be scattered throughout extracellular matrix, and unlike, for example, nerve cells or epithelial cells, make few or no contacts with each other.

A19–7.   (A) Collagen fibers. (B) Intermediate filaments. (C) Cellulose fibers.

A19–8.   Collagen is synthesized and secreted by the cell as individual polypeptide chains that are prevented from assembling into fibrils or fibers by peptides present at each end of the protein. These precursor procollagens are processed by proteolytic enzymes outside the cell, which remove the peptides from each end. The processed collagen chains then assemble into triple-stranded collagen fibrils, which then pack together into thicker collagen fibers.

A19–9.   D.

A19–10.   B.

A19–11.   E.

A19–12.   E.

A19–13. Their glycosaminoglycans are highly negatively charged and are thus hydrophilic. The multiple negative charges on the glycosaminoglycans are balanced by a cloud of positive ions such as Na^+, which in turn set up an osmotic potential that attracts water into the matrix, causing it to swell.

A19–14. 1. They provide a protective barrier around body cavities and on the outside of the animal. 2. They protect the interior of the organism from fluid loss. 3. They protect against invading microorganisms. 4. They allow the selective uptake of nutrients and release of waste materials. 5. They contain receptors for environmental signals.

A19–15. B.

A19–16. 1, apical surface; 2, tight junction; 3, adherens junction; 4, desmosome junction; 5, gap junction; 6, hemidesmosome junction; 7, basal surface.

A19–17. C.

A19–18. Add a water-soluble dye or radioactively labeled molecule as a tracer to one side of the epithelium, and examine its movement in the tissue. If the tracer stays on one side of the cell sheet, functional tight junctions are present.

A19–19. B.

A19–20. (A) They provide a strong mechanical bond between the cells. (B) They contain proteins (cadherins) that span the plasma membrane of one cell and bind via their extracellular portions to similar cadherins in the adjacent cell. The intracellular portions of the cadherin molecules are linked to the strong filaments (actin or intermediate filaments) of the cell's cytoskeleton, through which stresses are transmitted through the epithelium. (C) The hemidesmosome junction that attaches the epithelial cells to the basal lamina.

A19–21. C. Cyclic AMP is a small water-soluble molecule, and can therefore pass through gap junctions.

A19–22. A = 4, 6; B = 1, 3; C = 1, 5; D = 7; E = 1; F = 8; G = 4, 8; H = 1, 5; I = 1, 3.

A19–23. Cell lining the gut (few days) < epidermal cell (1 or 2 months) < red blood cell (4 months) < bone matrix (10 years) < nerve cell (lifetime).

A19–24. E.

A19–25. E.

A19–26. Since a mutation of this sort in the EGF receptor gene would lead to an increase in cell proliferation even if only one copy of the EGF gene were affected, this mutation is a dominant mutation. Mutations that deleted an EGF receptor gene would have either no effect or an inhibiting effect on cell division. Thus the EGF receptor is classed as a potential oncogene (a proto-oncogene).

A19–27. C.

A19–28. C.

A19–29. D.

A19–30.   These extremely rapid cell cycles consist of DNA replication (S phase) and a mitosis and cell-division phase (M phase) only, with no G_1 or G_2 phases. Thus the cells have practically no time to grow in mass before they divide, and so get smaller and smaller at each division.

A19–31.   B.

A19–32.   B. Although some cells (in the neural tube) behave as described in (E), other cells respond to Sonic hedgehog in quite different ways.

A19–33.   (A) Using a nucleic acid probe transcribed from the gene *in vitro*, you could use *in situ* hybridization to see the pattern of expression of the gene in the mouse embryo. (B) Having isolated the mouse gene, you could use recombinant DNA technology to make a "knock-out" transgenic mouse in which the mouse gene was inactivated or deleted. As a complementary test of gene function, you could make a transgenic mouse embryo by injecting a DNA construct, consisting of the gene coupled to a suitable control sequence, into a fertilized mouse egg, so as to obtain expression of the gene at abnormal sites, where its effects could be observed.

A19–34.   Hox genes are genes involved in recording and interpreting a cell's position in the embryo. Different Hox genes are expressed in different parts of the body. If, as a result of a mutation, a particular Hox gene is expressed in the "wrong" part of the body of a developing *Drosophila*, it will result in the cells of that body region producing structures characteristic of the part of the body where the Hox gene is normally expressed. The Hox gene involved in the *Antennapedia* mutation in *Drosophila* is one that is normally expressed in the part of the fly that produces legs. The mutation (in the control sequence of the gene) causes it to be expressed additionally in the head, in the cells that would normally form antennae. These cells therefore behave like leg cells, and form legs in place of antennae.